D0960115

ALIEN UNIVERSE

ALIEN UNIVERSE
Extraterrestrial Life in our Minds and in the Cosmos
DON LINCOLN

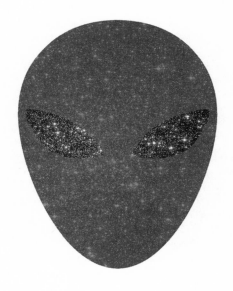

The Johns Hopkins University Press   Baltimore

© 2013 The Johns Hopkins University Press
All rights reserved. Published 2013
Printed in the United States of America on acid-free paper
9 8 7 6 5 4 3 2 1

The Johns Hopkins University Press
2715 North Charles Street
Baltimore, Maryland 21218-4363
www.press.jhu.edu

Library of Congress Cataloging-in-Publication Data

Lincoln, Don.
    Alien universe : extraterrestrial life in our minds and in the cosmos / Don
Lincoln.
        pages cm.
    Includes bibliographical references and index.
    ISBN-13: 978-1-4214-1072-2 (hardcover : alk. paper)
    ISBN-13: 978-1-4214-1073-9 (electronic)
    ISBN-10: 1-4214-1072-9 (hardcover : alk. paper)
    ISBN-10: 1-4214-1073-7 (electronic)
    1. Life on other planets.   2. Exobiology.   3. Extraterrestrial beings—Social
aspects.   4. Extraterrestrial beings in popular culture.   I. Title.
    QB54.L487 2013
    576.8'39—dc23   2012051094

A catalog record for this book is available from the British Library.

*Special discounts are available for bulk purchases of this book. For more information,
please contact Special Sales at 410-516-6936 or specialsales@press.jhu.edu.*

The Johns Hopkins University Press uses environmentally friendly book materials, including recycled text paper that is composed of at least 30 percent post-
consumer waste, whenever possible.

CONTENTS

ACKNOWLEDGMENTS

Books are never a solo effort. To convert an idea to a finished product takes the contributions of many.

Even the first idea can incorporate contributions from many people. This book wouldn't exist in its current form, were it not for my wife, who listened to my initial sketchy ideas and helped me weave them together into a coherent whole. The idea to interleave our evolving idea of Aliens with the concerns of the public at the time is entirely hers, and she also greatly clarified the presentation of several of the scientific topics I covered. Any residual rough spots in the execution are from when I stubbornly didn't listen to her.

Drs. Hassan Al-Ali, Albert Harrison, Judi Scheppler, Jill Tarter, and Erica Zahnle all provided expert advice on the text, correcting the misconceptions and errors. I am very grateful for their efforts and to Alvaro Amat for making helpful connections.

I would like to thank Patty Hedrick, Dr. Julie Dye, and Meredith Carlson for their assistance in procuring the drawings of Aliens, as imagined by children. I also thank both the Fermilab library and the Geneva Public Library staff for their invaluable assistance in finding original sources. The UFO literature is replete with fakes and manufactured "facts." To cut through various writers' agendas, it is imperative to confirm everything with original documents.

I am grateful to Linda Allewalt, Meredith Carlson, Sue Dumford, Vida Goldstein, Lori Haseltine, Dee Huie, Nancy Krasinski, Diane Lincoln, Toni Mueller, Robert Shaw, and Felicia Svoboda for being test readers. In addition to a myriad of typos, they helped identify weaknesses in the original manuscript that I hope I have removed. Michele Callaghan should get special mention for her deft copyediting. Finally, I would like to thank my publicist Kathy Alexander for her tireless efforts.

While these people all contributed directly to this book, it goes without saying that the information and tales recounted here wouldn't have been possible without generations of historians and scientists. Human knowledge is a cumulative endeavor, and what I've gathered here is really the work of hundreds of people.

It is traditional in my books to assign blame to a specific childhood friend of mine for any unnoticed residual errors. However it would be inappropriate to mention his name in a book that talks about Aliens, whom some think already walk among us. It wouldn't be right to blow his cover.

ALIEN UNIVERSE

THE QUESTION

The beginning is the most important part of any work.

Plato

One of the Biggest Questions of All

Of all the things that could change mankind's outlook, one of the biggest would be contact with intelligent life originating from another world. There have been similar paradigm-shifting episodes before in the history of humanity. When Galileo first saw the moons of Jupiter in about January of 1610, it marked the beginning of the end of the idea that Earth held a unique location in the cosmos. The publication of his manuscript *Starry Messenger* in Venice just two months later began the spread of the knowledge that someone had incontrovertible evidence that our planet could no longer be viewed as exceptional. The genie was out of the bottle.

Another paradigm-shifting event was the death of human exceptionalism, developed by Charles Darwin and his contemporaries. The idea that humans are but a twig on the vast tree of life forever changed our vision of ourselves as a unique species that had been given the Earth as our birthright. We became animals, like countless others, undoubtedly one with tremendous power to shape the globe, but still just a cousin species among many; one who had been shaped by the same forces that created the bear, the shark, and the kangaroo. Humanity's exceptionalism was reduced to our ongoing pride of our intellect and technical achievement.

While we know of other species on Earth who use tools and have a degree of intelligence, the discovery of no other species comparable to us has allowed some to continue to believe humanity to be exceptional and that it has been given, as the Bible says, dominion over all living things. Even the nonreligious might point to our spread from Africa across the globe, exploiting all environments and all niches, as a kind of Manifest Destiny of species. Taken to an extreme, some dream of a future in which mankind has left the Earth and spread across the cosmos, in a triumphant march of galactic conquest.

But what kind of galaxy will we encounter? Setting aside the very real practical difficulties associated with interstellar travel, is the galaxy a lifeless and sterile place, with the Earth being a unique and precious cradle of intelligent life? Or is the universe instead a cosmopolitan neighborhood, with many planets inhabited by species intellectually and technologically similar to our own? In short, we must ask a single important question.

"Are we alone?"

My purpose in writing this book is to explore that question. While imaginative thinkers have long speculated on the possibility and nature of life on other worlds, it was in the twentieth century that the idea has become common throughout Western culture. Unlike in early centuries, where discussions of alien life was the purview of the intellectual or academic, the idea of aliens has penetrated the literature, newspapers, movies, and other sources of information enjoyed by the public. The idea of Aliens, UFOs, alien contacts, abduction, interspecies reproduction, and so on, all exist to various degrees in the literature available to everyone.

This leads us to the fundamental question of the book. If Aliens exist, what do they look like? Or, even more specific still, what have members of the ordinary public *thought* Aliens look like over time? If you walked into a coffee shop in Greenwich Village in New York and asked the guy behind the counter what Aliens looked like, what would he say? What sort of answer would you get if you stopped instead for coffee in a diner in Ottumwa, Iowa? Or asked the same question 50 years ago?

Note the capital "A" in the word "Aliens." This choice is made to explicitly identify intelligent aliens, with which humanity could hypothetically one day compete for galactic domination. I am carefully distinguishing Aliens from alien life in general, which might be much more common in the universe. Aliens might one day appear in our skies and say "Take me to your leader," or they may attack the Earth for some of its resources. An alien duck flapping happily in the skies of a world circling Betelgeuse is *not* what I mean by an Alien. An Alien would be a sentient creature on that same world who has written a poem about that duck. An Alien must be intelligent, although not necessarily have extensive technology. An alien caveman counts as an Alien. An alien monkey does not.

Of course, the idea of alien life and Aliens go hand in hand, in the same way that life on Earth and humanity are inextricably linked. So, in parts of the book, we will occasionally expand the conversation to include a more generic discussion of alien life. But the focus will be on Aliens and how mankind's image of them has evolved and why.

Fiction versus Fact

So let's talk a little bit about the topics that will be covered in this book. To begin with, several different sources of information have guided our collective image of Aliens, which we can divide into three types: nonfiction, fiction, and a third category in which the boundaries between fiction and nonfiction are totally blurred.

Nonfiction comes from the best scientific thought of the era. Carl Sagan, of "billions and billions" fame, was an astrobiologist and a highly successful science popularizer. He and his colleagues have spent a considerable amount of time thinking about what our knowledge of the physics and chemistry of the physical world tells us about what Aliens might be like. More traditional biologists are finding life in more and more extreme environments, extending our appreciation of the versatility of life of the kind found here on Earth. However, not all planets in the universe are Earth-like, and it is possible that Aliens might be radically different from humans in the kind of air they breathe (if, indeed, they breathe at all), the temperature at which they thrive, the chemicals necessary for their metabolism, and so on. While science does not have a final answer on what an Alien might look like, a tremendous amount of progress has been made in understanding the range of the possible. We will certainly discuss these scientific subjects in this book, but the public's conception of Aliens tends to come not from the halls of academia, rather from the media and the entertainment industry.

"Space, the final frontier" are the opening words for one of the most successful science fiction television series of all time and the motivation for many a budding young scientist. In *Star Trek*, the crew of the starship *Enterprise* zips around the galaxy looking for "new life and new civilizations." The *Star Trek* universe consists of a grand diversity of species, where many worlds host life and indeed intelligent life. In fact, if Aliens didn't exist in the *Star Trek* universe, the show would have had a very different character. Without the Klingons, Cardassians, and Romulans (to name just a few), the humans on the starship would fly around the galaxy, poking at this barren rock or that, maybe occasionally encountering a nonsentient slime mold living on Epsilon Eridani IV. That wouldn't be as interesting as the social and political interactions that dominate the series's myriad plots.

Star Trek is not the only movie or television show that has shaped our thinking of Aliens. The bar scene in the first *Star Wars* movie showcases a diverse and colorful menagerie of Aliens, a "wretched hive of scum and villainy" as one character put it, socializing as humans would in the corner pub (or maybe a biker bar). In the subsequent *Star Wars* movies, many other creatures are introduced. Although Jabba the Hutt is an exception, most of the Aliens are vaguely humanoid, bipedal creatures with limbs and features that recognizably correspond to human analogs.

In fact, the bipedal structure of Aliens on both the big screen and little one has shaped the public's vision of them. In the old days, movie Aliens had to be bipedal because the actors that played them were obviously human. With the currently available techniques of computer graphics, movie makers no longer need to make such human-looking Aliens. However there is still the issue of making movie characters with whom the audience can relate. It's hard for me to imagine a successful movie that tells the story of star-crossed lovers with a species that involves three genders and has the color and consistency of lime flavored Jell-O. If you forgive the phrase, the storyline would be . . . well . . . just too *alien* to resonate with moviegoers.

This brings up a very important point. No matter how much science fiction buffs might devour the latest novel from the hot writer at the time, the size of the community of science fiction enthusiasts is relatively modest. Even a very popular Alien novel will reach a small number of readers. The written literature of science fiction has only affected the broader public in a brushing and indirect fashion. It is movies and television that have had the largest impact on the range of Aliens that is familiar to the public. In addition to the limitations of human actors and the need to provide a character with which the audience can relate, the stories of science fiction in the movies must be accessible to the audience. For instance, *Star Wars* has been described as a swashbuckling adventure, with a captured princess, a prince who didn't know his heritage, and an evil king. *Avatar* has been called *"Dances with Wolves* with blue people" and is considered to be a thinly veiled commentary on Western civilization's interaction with indigenous populations. And the movie *Alien* is similar to *Jaws* and any number of teen slasher movies. Science fiction movies are often an indirect commentary on the society and the politics of the time, just as George Orwell's *Animal Farm* is simply a metaphor for the Russian revolution (and, indeed, many a human revolution).

Many examples of fictional portrayals of Aliens reflect the cares that concerned humanity when the movie was released. The 1951 movie *The Day the Earth Stood Still*, in which an Alien and robot warn the Earth of the dangers of nuclear weapons, reflected the fears of post–World War II America. In a similar way, Edgar Rice Burroughs's stories of John Carter's adventures on Barsoom (i.e., Mars) in 1912 were clearly shaped by the last vestiges of postcolonialism. And even H. G. Wells's many stories reflect both the optimism and worries of the Victorian era.

Watch the Skies . . .

While we've briefly discussed the effect of both scientific thought and science fiction on the public's vision of Aliens, there remains a final, powerful influence on how the public views Aliens that is an indefinable mix of fact and fiction; a mystery wrapped in an enigma, with a blend of conspiracy and religious fervor tossed in to add spice to the tale. I am talking, of course, about UFOs.

Unidentified flying objects are also sometimes called flying saucers and are believed by some to be spaceships, ambassadors to Earth. Passions run high on this subject, ranging from those who believe we are not alone to others who believe the reports of UFOs come from a mix of charlatans, flakes, and well-meaning, but simply misguided, souls. People have reported seeing Alien craft, while others make the even grander claim of being in direct communication with Aliens. More recently some have reported being abducted by Aliens for various purposes, ranging from simple biological testing to interspecies reproduction. On the one hand, there is no doubt that some of the people who report these experiences believe them completely. On the other hand, it is also totally clear that the field is populated by hoaxes, fakes, and con men.

Although many, if not most, of the UFO/contactee/abductee reports are easily dismissed, some always remain unsolved. While "unsolved" doesn't mean "true Alien contact," the remaining air of mystery has certainly caught the attention of the public, the media, and even governments. Project Blue-book, administered by the U.S. Air Force, is merely the most well known of dozens of inquiries initiated by various government agencies into the UFO phenomenon.

The reports in the press of Alien encounters have an amplifying effect, with people seeing the reports and becoming susceptible to making additional (and confirming) reports. It's difficult to definitively explain what is going on. UFO believers will tell you that the increase in reported Alien encounters simply reflects a spike in Alien activity. Nonbelievers will tell you that any increase in the number of reports reflects a group delusion in the same way that a new report on the sighting of the Loch Ness monster or the Bigfoot will inevitably spawn more.

Regardless of where you come down on the question of Alien contact with humans, there is no disputing the fact that media reports of Alien contact beget more reports. Similarly they inform the public, science fiction writers, and filmmakers. The entertainment industry often then incorporates details of the reports into their stories. These tales of fiction then reach a larger au-dience and tell viewers what they should expect. This can induce additional reports, completing the cycle.

The point of this book is not to settle the question of (1) alien life, (2) the existence of intelligent Aliens, and (3) Alien visitation to the Earth. (Although I should probably state my opinion on those three subjects: (1) quite likely, (2) probable but very rare, and (3) exceedingly unlikely, respectively.) The point of the book is to discuss the prevailing vision of Aliens held by the gen-eral public both in the past and currently.

Figure P.1 shows some iconic Aliens, made famous by Hollywood and the media. All of these Aliens are recognizable to a majority of Americans, with the central figure being the one most commonly selected by adults when told to describe an Alien. As an exercise, I asked a large group of children, ages 4 to 11, to draw what they thought an Alien would look like. A sampling of these drawings is shown in figure P.2. These pictures were drawn indepen-dently, but there are striking similarities. Most of the Aliens are recognizably humanoid, with approximate bilateral symmetry. Those that do not have these properties are unlikely to be viable Aliens, as they apparently lack the ability to use tools. One striking similarity between the drawings is that the Aliens are all smiling and happy. Presumably this reflects responsible parenting and keeping the children from seeing too many scary movies. A few children have seen the archetypal "gray" Alien in the media: humanoid, with a high fore-

FIGURE P.1. Aliens are among us and have been for a long time. The iconic Aliens depicted here are familiar to anyone with even a modest knowledge of popular culture. See if you recognize them; their identities are given on the last page of the book.

FIGURE P.2. While adults have had many years to have learned what Aliens "should" look like, children are cleaner slates. Yet, as these images drawn by children show, some have already learned the "right" answer.

head, small chin, large almond-shaped eyes, and made famous by many an abduction tale.

Through the rest of this book, we will explore mankind's collective picture of Aliens. Chapter 1 will look at the concept of Aliens before 1900. This was the era where speculation about Aliens was generally the special province of scientists and theologians. Mars, being our nearest planetary neighbor, is a natural location to imagine where Aliens might exist, so I spend extra time in describing the rise and fall of the claims of intelligent Martian life.

In chapter 2, I describe "true" stories of Aliens: UFOs, contacts, and abductions. It is nearly irrelevant whether these tales are actually true. While some might object that this question is totally relevant, we must distinguish the question of alien life actually existing from the social phenomenon of Aliens as something embedded deeply in human culture. Aliens as pictured by mankind have their origins in stories in the media and entertainment industries, as well as tales told by a handful of people. Whether those tales are entirely true, a complete hoax, a misunderstanding of a natural phenomenon, or a manifestation of insanity doesn't matter at all. The stories and how they have moved through culture are what matters, and these stories have significantly shaped public opinion on the nature of Aliens.

Chapters 3 and 4 chronicle the evolution of Aliens in fiction in written literature, radio, television, and film. It is in fiction where authors can use Aliens in situations that are metaphors for the concerns of the society of the day. This is an especially interesting tale.

In chapter 5, we change gears. Rather than describing the historical opinion of Aliens, we use the rest of the book to explore modern efforts to understand what an Alien might look like. The first step in that process is to investigate what life on Earth can tell us. Chapter 5 surveys the various kingdoms of life on Earth, while chapter 6 ranges more broadly. Modern biochemistry and astrobiology has a lot to say about what kinds of life might exist "out there," including possible life-forms based on atoms other than carbon.

Chapter 7 completes our saga. In it, we move away from fiction and speculative science, instead focusing on the simple question: If you look for Aliens around nearby stars, what do you find? So far, despite half a century of looking and speculations that began even earlier, we have found nothing.

Until we find Aliens, we will continue to dream them. What we think they look like will tell us more about us than them. I don't know if we'll ever encounter extraterrestrial life. But, until we do, please join me and stare at the clear midnight sky and wonder.

ONE

BEGINNINGS

That the present inhabitants of Mars are a race superior to
ours is very probable.

Camille Flammarion

A floating silver saucer, perhaps punctuated with colorful lights. A diminutive gray being, with large, black, soul-less, almond-shaped eyes. Ghostly, telepathic voices. A hard, frigid slab. Silver medical instruments. Pokes and prods,
especially around the groin. Then a return to where you were, with an unease
and a period of time unaccounted for.

These are the elements of many a modern Alien tale.

For more than 70 years, humanity has slowly built a mythos around Aliens.
Even those of us who have no personal experience with UFOs, flying saucers,
or anything of the sort know the story. In this book, you'll learn from where
those elements have arisen. As we will see, that particular narrative is a recent one, built from a handful of progenitor tales and buttressed from being
told again and again both person to person and in the media. But, while the
general public's fascination with the question of extraterrestrial life has grown
tremendously in the past century or so, the interest isn't a new one. In this

chapter, you'll encounter scholars of the Renaissance who asked the question (and some who died for their temerity). You'll learn about ideas put forth in the nineteenth century, some in good faith and some just hoaxes to generate publicity. You'll learn about what our ancestors thought about our celestial neighbors: the moon and Mars.

And so we begin.

To discuss the existence of extraterrestrial life means, we must first answer a different question, specifically that of whether other planets exist. After all, if there are no other planets, it's hard to even ask the question of whether life exists on places other than the Earth.

The story starts, as it often does, with the early Greeks. Aristotle's writings had the longest impact on the question, and his argument was rooted in his physics and cosmology. For instance, Aristotle postulated a geocentric universe, in which the Earth was at the center, surrounded by a sphere of stars in fixed positions. Between the two were other spheres, each carrying the sun, the moon, and the wandering planets. These wandering planets were not imagined to be similar to Earth. Aristotle's physical theories postulated four elements: air, fire, earth, and water. He claimed that each of them had a natural affinity. Earth sank downward toward the planet, fire fled the planet, while water and air had intermediate affinities. According to his logic, this implied that there could be but one planet. Otherwise, earth wouldn't know where to fall . . . toward our planet or toward some other one. The logic was simple and the conclusion compelling. (It's also a scathing indictment of the role of pure logic in scientific discourse without empirical guidance.) While there were competitor ideas at the time, Aristotle's position dominated scholarly thinking for about 2,000 years.

If the question of extraterrestrial life hinged first on there being non-Terran planets, the first chink in the armor of Aristotelian logic can be traced to Nicolaus Copernicus. Just before his death in 1543, his book *On the Revolutions of the Celestial Spheres* was published. In it, he postulated a very different cosmology. In his heliocentric theory, the sun was at the center of the universe and all planets, including our own, revolved around it. And, of course, if the Earth is not central to the universe, then it is likely that neither is mankind. Copernicus did not write of the implications of his theory on the question of extraterrestrial life, but they were clear for others to pursue. Dominican friar Giordano Bruno, born just five years after Copernicus's death, was a bit of a Catholic bad boy. Eventually burned at the stake for religious heresies, he questioned many of the ideas accepted at the time. Relevant to our interests here, he postulated that if our sun was a star surrounded by planets, then all stars were suns surrounded by planets. If our planet held life, then others did too.

Galileo's *Sidereal Messenger*, published in 1610, further reduced the idea of Terran exceptionalism. He saw the moons of Jupiter and described the surface of the Earth's moon as having mountains and topography similar to the Earth. His contemporary Johannes Kepler was even more adventurous, suggesting that the moon was inhabited, with people living in caves on the side of craters. The Alien genie was out of the bottle.

The ensuing years involved discussions typical of the period between theologians, philosophers, and nascent scientists. In a period where the scientific instrumentation was not sufficient to settle the debate (a state of affairs that persists today), it is unsurprising that you would see the smart people of the era try to reason it out and many proposed hypotheses. There was no compelling winner in the debate over the question of whether other worlds carried life. We knew that there were other planets in our solar system and that other stars would most likely host their own planets. But, in a period in history in which learned people believed life came from a Creator, as opposed to natural processes, it is hard to imagine substantial progress being made on the question on the basis of reason alone.

Two important advances in scientific knowledge in the 1850s and 1860s put the discussion on more solid ground. First, Charles Darwin published his theory of evolution in 1859, which had as obvious an implication for extraterrestrial life as it did for the earthly variant. Secondly, the 1860s was the decade in which physicists started using spectroscopy in a serious way. Early spectroscopy used prisms to separate light into its constituent colors. For instance, studying the light absorbed or emitted by a gas allows scientists to determine its composition. In 1868, spectroscopic investigations of light emitted by the sun revealed a bright yellow line that couldn't be ascribed to the known elements, leading Sir Norman Lockyear to postulate that the sun contained an unknown element he called helium (after the Greek sun god Helios). Essentially, spectroscopy allowed scientists to do a chemical analysis without ever touching the object being studied.

In a similar way, scientists could turn their spectroscopes to light coming from planets in the solar system. By studying the spectrum, it is possible to ascertain the substances in the planetary atmosphere. Observation of oxygen, nitrogen, and water would indicate that the planet's atmosphere was like ours, where we know life exists. Combined with the knowledge we gain from evolution, it seems likely that life could form anywhere there was a favorable environment. It's not an airtight argument, but it certainly is a plausible one and one we will return to toward the end of the book. The mid- to late-1800s marks the point where answers to the question of extraterrestrial life became accessible through the mastery of scientific instruments.

By this period, telescopes were good enough to be able to study the moon's surface in detail. It was clear to all but a few eccentrics that it was a lifeless ball, or at least so it appeared. No water, no atmosphere, nothing but rocks and craters. With the moon out of the picture, the scientist's attention turned to Mars and Venus, as they were our planetary neighbors. In a later chapter, we'll again see this fascination with the neighbors in our study of Aliens in science fiction.

1835 Moon Hoax

Before we continue our story of mankind's search for Alien life among nearby planets, we must recall that this book is not just about what scientists think and thought, but also about what the public thinks. Prior to the ability of science to totally debunk the idea, the possibility of lunar life was seen as plausible. A set of stories in the New York *Sun* in August 1835 brought Aliens to their readers in a dramatic and splashy way.

To better understand the tale requires going back in time about five years before it begins and taking a look at early nineteenth-century journalism. In 1830, newspapers were different from the ones we have now. There were typically only two types of newspapers in that era: political ones and business ones. The political ones were published by political parties to advance their specific agenda, while the business ones were written for the business community to inform the affluent about what was going on in the economic sphere. Modern-day equivalents of the latter might be the *Wall Street Journal* or the *Financial Times*. Newspapers were sold by subscription and cost six cents a day or about twenty dollars a year. That was a fair bit of money at that time, and, consequently, newspapers tended to be read by the well to do and might have a circulation of one or two hundred readers. The newspapers were conservative, in that they tended to stand behind the material in their pages. (Although their politics might be not be conservative, indeed they could be rather radical.) In a way, carrying an advertisement was an endorsement.

The world changed on September 3, 1833, when Benjamin Day began publishing the New York *Sun*. Perhaps the most famous story written in the *Sun* was the 1897 editorial "Is There a Santa Claus?" (most commonly called

"Yes, Virginia, There Is a Santa Claus"). However, in 1833, the *Sun* was a game changer, as it was sold for a penny per copy. It was the first of the newspapers in New York City that became what was known as the "penny press." Because the cost was lower, the only way newspapers like it could stay in business was through volume sales. The phrase "Extra, extra, read all about it" stemmed from this time. In the months prior to the story I'm about to tell, the daily circulation of the New York *Sun* had reached about 20,000 copies. The penny presses were closer to what we currently call tabloids, filled with hearsay and stories from the police blotter, full of salacious details. If they carried an advertisement, it certainly didn't imply an endorsement. The readers expected to be entertained as well as informed. And, as we shall see, it is from such a periodical that one of the first media frenzies came to be. On Friday, August 21, 1835, the *Sun* published a small teaser notice on the second page of the paper: "We have just learnt from an eminent publisher in this city that Sir John Herschel at the Cape of Good Hope has made some astronomical discoveries of the most wonderful description, by means of an immense telescope of an entirely new principle."

Sir John Herschel was an excellent scientist and mathematician. Son of Sir William Herschel (discoverer of the planet Uranus), he built a telescope with a diameter of 18 inches and a 20-foot focal length that allowed him to explore the heavens in great detail. For his scientific work, he was made a Knight of the Royal Guelphic Order in 1831. He left England for South Africa in the fall of 1834, bringing his telescope with him. The goal was to study the southern sky.

Given Hershel's reputation, it is perhaps unsurprising to see an announcement of his work if he had made an advancement in astronomical instrumentation. The public of 1835 was just as fascinated by the heavens as we are today. Other papers in New York made no mention of the announcement.

On Tuesday, August 25, the *Sun* began publishing a series of columns over six days describing observation of life on the surface of the moon. And not just ordinary forms of life were observed, but rather intelligent life with an advanced civilization. However, the first day was a little more ordinary. It described a new telescope. The series of columns was entitled *Great Astronomical Discoveries Lately Made* by Sir John Herschel and was supposed to be a reprint from a supplement to the *Edinburgh Journal of Science*. In essence, this was as if the newspaper was reprinting a special issue of a Scottish scientific journal, although the editor told the readership that some technical and mathematical details had been omitted. The newspaper article was accompanied by an editorial note that said, "We this morning commence the publication of a series of extracts from the new Supplement to the *Edinburgh Journal of Science*, which have been very politely furnished us by a medical gentleman immediately from Scotland, in consequence of a paragraph which appeared on Friday last from the *Edinburgh Courant*. The portion which we publish today is introductory to celestial discoveries of higher and more universal interest than any, in any science yet known to the human race." As it happened, the *Edinburgh Journal of Science* had suspended publication two years earlier, but that was not widely known.

The first day described a new telescope, with a lens of 24 feet in diameter, made of excellent glass. The weight of the lens was a little over seven tons. Mind you, the biggest telescope ever built using a lens (rather than a mirror) had a diameter of 49 inches. But the telescope became even more outlandish. Because of its great size, it was capable of even studying "the entomology of the moon, in case she contained insects upon her surface." That's a pretty impressive claim. In addition to the large telescope, the superb performance was made possible by the use of a "hydro-oxygen microscope" to brighten the image. Essentially, the claim was that the telescope fed into a microscope and thus the ability to closely study the surface of the moon was achieved.

If you read the original article, you are struck by the presence of many details that make it sound more authentic, like the manufacturer of the lens, the name of Herschel's assistant, and the assistant's relationship with Herschel's famous father. Nowadays, this attention to detail sounds like the output of a gifted and diligent investigative reporter. However, as we will see, it was instead a delightful tall tale, told with enough detail to convince many a reader.

Day two of the saga began with a discussion of why the telescope needed to be placed in the southern hemisphere, but it finally got down to brass tacks and described what Herschel saw as he peered at the surface of the moon or, as the article stated, "no longer withholding from our readers the more generally and highly interesting discoveries which were made in the lunar world." What did he see?

Well, the first thing observed was basaltic rock, but as the Earth turned, what moved into his field of vision was a rock shelf "profusely covered with a dark red flower," similar to rose poppies seen on earthly cornfields. Further inspection revealed trees, but only one kind, large and reminiscent of yew trees on Earth. Alien life had been observed but only of the plant variety.

Further searching revealed beautiful crystals, huge and colored vibrant purple and vermillion. Landscapes beyond imagining and a vast forest, this time with trees "of every imaginable kind," the author reported continuous herds of brown quadrupeds that looked very much like bison. The bison were followed by gregarious, "bluish-lead" unicorn-goats. Pelicans, cranes, a strange, spherical, amphibious creature that rolled along the beaches—animal life had been observed.

The article on day three spoke of more geology and the first observation of intelligent, although primitive, lunar life. This life took the form of a bipedal, tailless, beaver that carried its young in its arms and lived in small huts. Smoke in the vicinity of the huts revealed that the beavers had conquered fire. According to the article, the question of intelligent extraterrestrial life had been definitively answered, although the best was still to come.

Day four was perhaps the high point of the narrative, when intelligent humanoids were observed. They were about four feet in height and covered with short and glossy copper-colored hair, except on their faces. Their faces were yellowish, similar to an orangutan. They also had wings. The wings were bat-like, and so the author named them Vespertilio-Homo (or bat man). While the observers watched the creatures' behavior, the article deferred a discussion of what they saw for a later and more detailed article. Mankind was no longer alone in the universe.

It would be hard for day five to eclipse the revelations of the day before. Literary custom required a denouement. The article discussed more geology, observations of oceans, islands, and so forth. However a particular valley stood out with hills built of snow-white marble or perhaps semi-transparent crystal and adjacent to a flaming mountain, for in this valley stood what appeared to be an abandoned temple, triangular in shape and made of pure sapphire. The roof was made of a yellow metal, flame-like in construction. Since the temple seemed abandoned and all the observers saw were flights of lunar doves landing on the pinnacles of the roof, they were unable to speculate on the meaning of the temple's imagery. Further searching revealed two other temples located a distance away.

Day six was the final installment in the saga of Vespertilio-Homo. The astronomers saw more bat people, this time closer to the temples. These bat people were bigger than the former ones, lighter in color and "in every respect an improved variety of the race." Happy and social, these new people sat around in groups passing the time, "We had no opportunity of seeing them actually engaged in any work of industry or art; and so far as we could judge, they spent their happy hours in collecting various fruits in the woods, in eating, flying, bathing, and loitering about on the summits of precipices." With these observations, the record ends of the study of Vespertilio-Homo.

The article goes on to say that the astronomers left the telescope and went to bed, only to be awakened the next day to find that the telescope had inadvertently lined up with the sun and the resultant image started the building on fire. Luckily no serious damage had been done, but it took several days for the soot and mess to be cleaned up, by which time the moon was no longer found in the night sky. Herschel then turned to studying the rings of Saturn, which he found to be debris of two worlds that had collided.

Herschel was busy cataloging his observations of stars he had seen, so his assistants looked again at the moon, this time seeing an even superior form of Vespertilio-Homo. "They were of infinitely greater personal beauty, and appeared in our eyes scarcely less lovely than the general representations of angels by the more imaginative schools of painters." The author (one of Herschel's assistants) then closed by saying that he would defer discussion of these angelic bat people until Herschel could write something himself.

Needless to say, the real Herschel had no part in this. He actually was doing research south of the equator and was more than a bit aggrieved when he later heard the liberties that had been taken with his reputation.

So ended the six columns in the *Sun*. What impact did these columns have on the public? Well, quite simply, it was huge. The *Sun* sold out of its total circulation of about 20,000. Further, the competing papers in New York reprinted the story. Approximately 100,000 copies of the article were printed in New York City alone (at a time in which the population of New York was only about 300,000). With no radio, nor even telegraph, the story travelled across the country relatively slowly, although it arrived in the other major Eastern cities like Boston, Philadelphia, and Baltimore in a matter of days. It took a couple of weeks to make it to the Midwest and a month to Europe. English and French journals reprinted the articles, without naming the source of the material as a penny press newspaper in the United States. The story was even reprinted in Edinburgh. Given that the *Sun* attributed the original source to the *Edinburgh Courant*, presumably the Scottish people knew it was a fake, but they reprinted it anyway.

While the *Sun*'s circulation didn't change dramatically with the story, they did print a pamphlet containing all six columns, accompanied by several lithographs showing imagined renditions of Herschel's discoveries. One of these is reproduced in figure 1.1. While the *Sun* never disclosed how many pamphlets

FIGURE 1.1. This lithograph did not appear in the newspapers but was included in a subsequent pamphlet printed by the New York *Sun,* which included all six articles detailing the moon hoax of 1835, as well as several figures that gave dramatic emphasis to the text. New York *Sun.*

were sold, later writers estimated the number to be about 60,000. At a cost of twelve cents a pamphlet, the *Sun* did end up making a fair bit of money.

With 100,000 copies of the story printed in New York, along with a very large number printed elsewhere across the United States and the world, the moon hoax of 1835 was one of the first media events and something that would have been impossible just five years prior. The invention of steam-powered presses, along with less expensive paper, made it economical to produce newspapers in great quantities. When this was combined with the business model that sold the papers for a penny and utilized, for the first time, newsboys on street corners to sell them, large numbers of people could be reached quickly. The hoax also had an impact on journalism as a whole, starting a discussion on the question of standards in journalism and whether reporters had an obligation to report the truth.

It was not very long before the *Sun*'s report was generally conceded to be a hoax, but there was a brief period of time when the broad public was transfixed by the idea of extraterrestrial life. The scholars of the day still debated the question of whether the moon might host life. The evidence was rather strongly against, most notably by watching the moon pass in front of stars. The image of the stars in telescopes remained crisp until the very last second, suggesting that the moon had no atmosphere. Air on the moon would have made the stars' images blurry. However, while the debate continued among the community of scholars, the question was less mentioned among the public. The *Sun's* story brought it to the fore. Thinking about Aliens was now mainstream.

Mars

While the moon hoax of 1835 was entirely a fictional event, the question of life on Mars remained scientifically reputable for a much longer time. For one thing, Mars is much farther from Earth, making it much more difficult to image. In addition, the diameter of Mars is twice as large as the moon, making the planet more Earth-like. Polar ice caps were observed on Mars as early as the middle of the seventeenth century and studied in some detail by William Herschel (the father of John Herschel of moon hoax fame). In fact, speculation on the question of life on Mars (especially intelligent life) reached a fevered pitch in the late 1800s.

Perhaps the best place to start our story is with French astronomer Camille Flammarion. He was a popularizer of science, although his readers were equally likely to be academics as members of the educated general public. His first book *La Pluralité des mondes habités* (*The Plurality of Inhabited Worlds*) was published in 1862 and put forth the idea that there were many inhabited worlds in the universe. He was not the first to suggest the idea, but he was among the first to suggest that extraterrestrials might be truly alien, as opposed to mere variants on humans. In two of his books, he proposed several exotic species, including sentient plants.

His book *Astronomie populaire* was published in 1880 and translated into the English *Popular Astronomy* in 1894. The book is full of speculation about extraterrestrial life, both lunar and Martian, and sold more than 100,000 copies in French. His 1892 book *La Planète Mars et ses conditions d'habitabilité* (The Planet Mars and Its Conditions for Life) supported the idea of Martian canals built by an advanced civilization.

Flammarion was not the originator of the idea of Martian canals. That distinction came from Italian scientist Giovanni Schiaparelli. And to understand that tale requires that we learn some basic astronomy.

The orbital period of Mars is 687 Earth days, and its orbit is also highly eccentric, ranging from 129 million miles to 155 million miles from the sun. Consequently, about every two years Mars and the Earth are relatively close at opposition. This term means that Mars was opposite the sun and thus could be seen directly overhead at midnight. When the Earth's orbit is taken into account, about every 15 years the two planets are especially close. Because of these astronomical factors, the years 1877, 1892, and 1909 were especially auspicious for viewing Mars, as it appeared to be about twice as wide than it did in more pedestrian years.

While astronomers had watched Mars for millennia, it was in 1877 that the Martian chronicles heated up, for that was the year that Giovanni Schiaparelli reported observing "canali" on Mars. Canali is an Italian word that means "channels," but it was mistranslated into English as "canals." And "canals" has an important implication. It means an artificially dug water course. In an era when the Suez Canal had recently opened (1869) and the digging of the Panama Canal had begun (1881), it is inevitable that the word would excite the imaginations of people who heard it. In the 15 years between the oppositions of 1877 and 1892, there was speculation on the nature of the canals and even bitter disagreement on whether they existed at all. The telescopes of the day were typically refractors and consequently were relatively small. It was rather difficult to clearly resolve features on Mars, and so the question of whether canals were observed was necessarily a subjective one. While observations in subsequent years were not under the optimal conditions of 1877, other astronomers also reported seeing canals at observatories across the world. Others didn't and the debate raged within the astronomical community.

The question of artificial canals on Mars was a pressing one, and astronomers looked forward to the next optimal opposition in 1892 to hopefully resolve the issue. Camille Flammarion's 1892 book on Martian habitation and Schiaparelli's 1893 *La vita sul pianeta Marte*—which literally meant *Life on the Planet Mars*—were timely. A budding scientist's receipt of Flammarion's book as a Christmas present had an unanticipated large impact on the debate on Martian canals and the awareness of the public of the question.

Percival Lawrence Lowell was born into an affluent family in Boston, Massachusetts, on March 13, 1853. His family made their money in the Lowell textile industry, and he was a sixth-generation student at Harvard University. Lowell was a brilliant student and interested in science. At his college graduation in 1876, he gave a speech on the "Nebular Hypothesis" describing the formation of the solar system. After graduation and the obligatory tour of Europe, Lowell attended to his family's business affairs and traveled extensively in the Far East, where he wrote several books about Japan that were well received back in the United States.

In 1893, the year in which Schiaparelli's provocatively named book came out, Lowell received the present that brought little green men to the public. After devouring Flammarion's book, he decided to become a full-time astronomer and to focus on the planet Mars. In mid-January of 1894, the Boston papers reported that Lowell had decided to finance an observatory in Arizona. The location was selected because of its altitude and dark and clear skies. Flagstaff, Arizona, became the center of Martian research.

Research began rapidly, for if they missed the 1894 opposition, the next favorable opposition was 15 years in the future. Lowell's observatory went up quickly, and he turned his telescopes to Mars. Initially, he and his team used two temporary telescopes, one 12 inches and one 18 inches. He saw canals and lots of them. Eventually 183 canals would be reported by Lowell and his associates; the first article came out in late summer of 1894 (figure 1.2).

This article not only described the canals he observed, but it went much further, revealing his underlying motivation. For, while traditional astronomers might want to understand Mars, it was clear that Lowell had already made up his mind. He was confident that he was seeing the signature of Martian civilization. Mars was thought to be an older and dying world, dry and increasingly desolate. He believed that the ancient civilization of Mars had built a vast network of canals to bring water from the polar ice caps to the mid-latitude and equatorial areas in an attempt to survive. Dark patches that were observed in the telescopes were thought to be oases in which pockets of Martians continued to eke out a harsh and desperate existence. Lowell detailed his ideas in three books: *Mars* (1895), *Mars and Its Canals* (1907), and *Mars as the Abode of Life* (1908).

Lowell was not merely an amateur astronomer. He was a scion of a wealthy Boston family, charming when he wanted to be, and passionate about his interests. In addition to observing the heavens, Lowell moved in fashionable circles. His name and wealth gave him access to the movers and shakers of the day. He was invited to the "A-List" parties, where he would dazzle the attendees with his ideas on Mars. Publishers of newspapers and magazines who were in attendance knew a good story when they heard one. The stories showed up in print. Lots.

Lowell has been aptly called the most influential popularizer of astronomy before Carl Sagan. Stories about him were splashed conspicuously across leading periodicals. For instance, on December 9, 1906, the Sunday edition of

FIGURE 1.2. Percival Lowell and his assistants catalogued many canals they believed they had observed on the surface of Mars. This 1905 drawing gives an indication to the extensive canal network he thought he found. Courtesy of the Lowell Observatory Archives.

the *New York Times* ran a column about Lowell that took up well over 80% of the front page, with the title "There Is Life on the Planet Mars." The author was rather taken by Lowell: "This discovery is due to the brilliant genius, the persistent energy, and the marvelous power in research of Percival Lowell."

While Lowell's fame in the popular press was high, there were many doubters in the scientific community. The situation was not one in which there were but two positions: canals and no canals. Some astronomers accepted canals, but as natural phenomena, while others accepted splotchy features on the Martian surface that changed over time and were taken to be seasonal vegetation variation. Astronomer W. W. Campbell reviewed Lowell's book *Mars* and said, "Mr. Lowell went direct from the lecture-hall to his observatory in Arizona; and how well his observations established his preobservational views is told in his book." Campbell accepted the canals as real features, but he found ridiculous the attribution of the features as evidence of intelligent handiwork. Campbell was also aware that the amount of water available in the atmosphere of Mars was exceedingly low and found the lack of water compelling evidence that there could be no civilization on the planet.

The impact of Lowell's advocacy can be measured in many ways, but perhaps the strongest is the appearance of stories of Martian civilization in fiction. Possibly the first occurrence would be H. G. Wells's 1898 novel *War of the Worlds*. During the late 1880s, Wells was trained as a science teacher, and he had written a biology textbook. However, in 1894 he joined the scientific journal *Nature* as a reviewer. Much of his writing served to translate the highly technical innovations of the Victorian era into terms familiar to the educated lay reader. His essay *Intelligence on Mars*, published in 1896 in the *Saturday Review*, speculated about life on Mars and how the inhabitants would cope with what he considered to be an older planet. Much of the article, including his conjecture that the Martians might move to another planet to survive, was found in his famous fictional work *The War of the Worlds*. He even incorporates the reports of a flash of light observed on Mars by an astronomer in 1894 (and published in the August issue of *Nature*) as the start of the book. As will be detailed in chapter 3, *The War of the Worlds* describes the invasion of Earth by Martians and their subsequent defeat by Earth microbes.

Lowell is a central figure to the excitement about Martian intelligence, but he was neither the originator of the idea, nor did he resolve it. He was merely a true believer, articulate and enthusiastic, who excelled at communicating his vision. Indeed, Lowell never did really give up his beliefs, even when they were ruled out by better measurements.

The year 1909 was when there was another particularly favorable opposition for Mars and when Martian canals were ruled out, at least as far as the scientific community was concerned. The scientist who dashed the dreams of those who hoped it had been proven that mankind was not alone in the universe was Eugene Antoniadi, a Greek astronomer who gained some fame in later life as a scholar of ancient Greek and Egyptian astronomy. That Antoniadi was the one who resolved the debate came with some irony, as he worked at Flammarion's observatory in 1894 and published his results in the journal of the French Astronomical Society, which Flammarion began. But such is the small world of professional astronomy.

Antoniadi was able to see dark, irregularly shaped spots on the surface of Mars, but he concluded definitively that the canals themselves were "an optical illusion." His result made it to the United States, where a new class of telescopes was coming online, the big reflectors. The 60-inch reflector on Mount Wilson was turned to Mars, and the director wrote to Antoniadi, saying, "I am thus inclined to agree with you in your opinion . . . that the so-called 'canals' of Schiaparelli are made up of small irregular dark regions." Antoniadi continued to observe Mars, writing his own book *La planète Mars* in 1930. But in 1909 the astronomical world moved on.

As is often the case in these situations, there were true believers who refused to accept the new conclusions. Until his death in 1916, Lowell maintained that those who failed to see the canals were mistaken and doing sloppy work. Further, he still had the ear of many of the leaders in the popular media. For instance, in the August 27, 1911, issue of the *New York Times* Sunday magazine, a splashy article entitled "Martians Build Two Immense Canals in Two Years" described two canals, each a thousand miles long and 20 miles wide that had appeared on the Martian surface. The possibility that these were natural features was ruled out in the article.

The public was not as quick to give up on Martian canals as was the scientific community. First, they were not as close to the data as the astronomers were and, second, they had received a steady barrage of stories, speculating on Martian culture and how the civilization must be frantically trying to save itself. It was a gripping saga and not one that can easily be forgotten. Edgar Rice Burroughs's Barsoom saga (Barsoom being Burrough's Martian name for Mars) began in 1912 with *A Princess of Mars*, and we will learn more of this iconic series of stories in chapter 3.

Wrap Up

The idea that we are fellow travelers in this universe is not a new one. As we have seen here and can be followed up in the suggested reading, there have been centuries of arguments over the question of extraterrestrial life; they were theological, philosophical, and quasi-scientific. However it wasn't until the last years of the nineteenth century that the thought of life of non-earthly origins became a common topic of conversation outside the circles of the highly educated.

The reasons for the broader dissemination are various. First, the scientific instrumentation became better, allowing for more definitive arguments among academics. After all, questions like the existence of extraterrestrial life or intelligence is an empirical one, and there is no chance that a theological or philosophical discussion will definitively resolve the debate. Improvements in telescopes and the new technique of spectroscopy allowed for solid discussions, well informed by hard data. However the improved science doesn't explain the change in the level to which the public was informed. For this, you need a communication method. In the 1800s, improvements were made in printing technology and the way in which the printed material was brought to the public. Technology made it much easier for people to learn about the sorts of things that interested them, as evidenced in the tremendous response to the moon hoax.

As we will see in chapter 3, the first half of the twentieth century showed an increase of what we now call science fiction. While stories of extraterrestrials are not the only tales written in that genre, the Alien ones became somehow respectable, given the vast number of newspaper articles people had read about Mars. This is not to say that our version of Aliens hasn't evolved since the first decade of the twentieth century. Indeed, our current view of Aliens differs dramatically from the speculations of Lowell, Wells, and their contemporaries. To understand how that came to be, we must turn to a world convulsed in war.

TWO

ENCOUNTERS

> Deep Throat: Mr. Mulder, why are those like yourself who
> believe in the existence of extraterrestrial life on this Earth
> not dissuaded by all the evidence to the contrary?
> Mulder: Because all the evidence to the contrary is not entirely
> dissuasive.
> Deep Throat: Precisely.
> Mulder: They're here, aren't they?
> Deep Throat: Mr. Mulder, they've been here for a long time.
>
> *X-Files*, Season 1, Episode 2

The *X-Files* was a highly successful science fiction television show that ran from 1993 to 2002. In it, two FBI agents, Fox Mulder and Dana Scully, are tasked with investigating odd reports that are stored in the classified "X-Files." While something like two-thirds of the episodes were devoted to the "monster of the week" (e.g., investigating whether a vampire or werewolf was involved in a string of murders), the remaining episodes were used to develop a story-line about Aliens on earth and the government's cover-up of what they know.

This television show is an excellent example of how the media, entertainment industry, UFO devotees, and people who claim to have been abducted by extraterrestrials have interacted with one another and shaped one another's views. Fact (meaning honestly believed reports of UFO sightings and abductions) and fiction are inextricably interwoven, leading to a narrative that is well known to society. A 2008 poll showed that 36% of Americans believe the Earth has been visited by Aliens and that 80% think the government knows more than it's telling. Ask a random stranger (which I've been doing lately, generating some peculiar looks) what Aliens look like and what happens if you are abducted by them, and you get stories that are broadly similar; short, gray humanoids, with huge foreheads, small chins, and pupil-less black eyes. Further, the Aliens are inexplicably fascinated with the human reproductive system, probing it with various silver-colored implements. How can people with the most miniscule interest in Aliens be so aware of the abduction narrative? That kind of penetration of the culture takes years. In the following chapters, we'll take some time to explore how that story developed and was disseminated.

We've talked a bit about earlier media and public interest in the moon, Mars, and Martians, but it was the 1940s where our tale of Alien contact began to take off. As we move forward, we need to keep a very important thing in mind. Students of UFO-ology have at their disposal an enormous literature to read. Tens of thousands of tales of "real" Alien contact have resulted in hundreds of books and many websites. Governments around the world have launched dozens of inquiries into the question of Alien visitation. Anyone who wants to immerse themselves in the literature of this culture has a daunting task before them. But we're not going to do that.

In this book, we're not interested in this obscure sighting or that unexplained abduction tale. We are interested instead in the "big" stories, the ones that got a lot of publicity, for only the ones that had extensive (and ongoing) media coverage are able to enter into the public consciousness. It will likely not surprise you that many elements of the stories that people tell about their contact with Aliens were already present in fiction accounts, which we look at closely in chapters 3 and 4. However, our current concern is to understand how a flight by a solo pilot in the 1940s or a long drive by an interracial couple in the early 1960s could change our collective vision of extraterrestrial life. Our tale begins in earnest over the skies of Europe, as the Allies tried to push the Nazi armies back into Germany.

Foo Fighters

Carl von Clausewitz wrote in his book *On War*, "The great uncertainty of all data in war is a peculiar difficulty, because all action must, to a certain extent, be planned in a mere twilight, which in addition not infrequently—like the effect of a fog or moonlight—gives to things exaggerated dimensions and unnatural appearance." He was writing on the difficulty for commanders to get full situational awareness and its effect on their subsequent decision making. But war is an adrenaline-raising situation that has an effect on a combatant's perception. Incomplete information, conflicting reports, and high stress mean that mistakes will be made.

Let's face it. Being in a B-17 over the skies of Germany between 1943 and 1945 pretty much guaranteed that you would be a little tense. Something about the strafing by the Luftwaffe and tons of antiaircraft flak objecting to your visit is bound to add a little excitement to your day. I imagine that a pilot in a P-51 Mustang flying combat air patrol and tagging along for the ride probably shared in the bombardiers' heart-pounding experience.

These are the kinds of men who reported what is now generally accepted as the first observations of the phenomenon that would later be called flying saucers. Airmen over Europe began making reports of seeing balls of light that shadowed their airplanes as they flew through the skies. The balls of light would cling to wing tips, even when a pilot pushed his fighter over in a dive that approached speeds of 360 mph. Other balls of light would tail them or travel in parallel paths but not be in contact with the plane. Occasionally, a pilot could outrun the lights. "Kraut fireballs" or "foo fighters," as they became known, were not seen as potential extraterrestrials, but rather in terms of possible Nazi weapons to explain and counter.

A report in the January 2, 1945, issue of the *New York Times* quoted a pilot as saying, "There are three kinds of these lights we call 'foo-fighters.' One is red balls of fire which appear off our wing tips and fly along with us; the second is a vertical row of three balls of fire which fly in front of us and the third is a group of about fifteen lights which appear off in the distance—like a Christmas tree up in the air—and flicker on and off."

The report goes on to state that foo fighters were thought to be German in origin and both a psychological as well as a military weapon, although "it is not the nature of the fire-balls to attack planes." A second pilot thought at first that they were "some new form of jet-propulsion plane after us. But we were very close to them and none of us saw any structure on the fire balls."

This report was not the only one. An Associated Press report from Paris two weeks earlier (December 13, 1944) said that the Germans had thrown silvery balls against pilots doing daytime bombings and that these balls appeared both individually and in clusters. This report was repeated in the January 15, 1945, issue of *Time* magazine. However, this article showed that the reports of foo fighters did encounter skepticism. Some scientists dismissed the balls as persistent visions induced by pilots seeing flak bursts. Others suggested St. Elmo's fire or ball lightning.

It is interesting to read what kinds of wilder speculation appeared in the press. In the *Time* magazine article, "front-line correspondents and armchair experts had a Buck Rogers field day," guessing that the balls of fire were a weapon remotely controlled via radio (which was dismissed as being absurd, given that the balls exactly tracked some plane's movements), as well as other prosaic phenomenon. A few more ideas that were kicked around were that the foo fighters were intended to (1) dazzle pilots, (2) serve as aiming points for antiaircraft gunners, (3) interfere with the plane's radar, or (4) interfere with the operation of the plane's motor, perhaps stopping the plane in midair. But, in the context of this book, which is the view of Aliens in the public eye, it is relevant that extraterrestrial origin wasn't one of the suggestions. Modern UFO enthusiasts point to foo fighters as the first hints of Alien contact, but this wasn't in the minds of the people reporting bright lights in the sky. They had a war to fight. But the idea of Aliens was about to begin.

UFO 1947

June 24, 1947, was a turning point in what we collectively think when we turn our eyes to the night sky. Kenneth Arnold was a businessman and a pilot. He never flew a bomber or fighter above Europe, but he certainly encountered men in the hangar who had. Arnold was flying his private plane near Mount Rainier in Washington State when he reported seeing nine brilliantly lit objects, flying across the face of Rainier. He described them as being flat, like a pie pan, and thin enough that they were hard to see. They were sort of half-moon shaped, convex in the rear and oval in the front (figure 2.1). The objects moved independently, but in a line, like the tail of a kite.

Arnold was flying at 9,200 feet at a speed of about 115 mph. He judged the objects as being at about 10,000 feet and estimated their speed at about 1,800 mph, although he allowed that there might be a mistake in his estimates and so he stated 1,200 mph was a more reasonable guess.

When he landed in Yakima, Washington, Arnold told the manager of the

FIGURE 2.1. This picture appears in Kenneth Arnold's *The Coming of the Saucers* to give an idea of what he saw in 1947. You will note that the term "flying saucer" does not accurately describe this shape. Copyright Ray Palmer.

airport, who didn't believe him. While in Yakima, Arnold talked to other people who happened to be at the airport. Arnold then flew on to Pendleton, Oregon, where an air show was going on. He was unaware that someone from Yakima had called ahead and told people there that he had seen something odd flying in the air of southern Washington.

In Pendleton, he told his story to aviator friends, who weren't surprised, nor did they discount it. First, Arnold was known to be a man of excellent character, and, second, some of the pilots had heard similar stories while flying sorties over occupied Europe. Whether foo fighters or some new airplane being tested by the Army Air Force, the observation was a curiosity but not something to get incredibly worked up about. The most noteworthy factor, and perhaps the reason that the press became involved, was the speed that he quoted for the aircraft. Going 1,200 mph was awful fast for 1947. It's pretty fast even for today.

It wasn't until the next day that Arnold spoke to reporters, when he stopped by the offices of the *East Oregonian*, a newspaper in Pendleton. He told them his story, which they regarded as plausible enough to publish and put on the news wire for further dissemination. And that, as they say, was when things got crazy. The story was picked up by United Press International and the Associated Press. Some big newspapers carried the story. The *Chicago Tribune* ran a story two days later on page one, titled "See Mystery Aerial 'Train' 5 Miles Long." However there was no mention of UFOs or flying saucers. The report quoted Arnold as saying they were fast, reflective, and moved like the tail of a Chinese kite, as if the craft were connected by a string. It further reported that the Army wasn't doing high-speed tests in the area.

The term "flying saucer" seems to have been coined by accident. Arnold told reporters that the nine objects he saw were flat and shiny like a pie pan and that they looked like a little fish flipping in the sun. On June 26 the Chicago *Sun* ran an article "Supersonic Flying Saucers Sighted by Idaho Pilot." However, this seems to be an addition by an editor or headline writer. Much later, Arnold recalled that he had told the first reporters he spoke with that "they flew erratic, like a saucer if you skip it across the water," and this phrase seemed to turn into "flying saucer," which was then used and reused by newspapers. However, in the early newspaper articles, Arnold is never quoted as saying the phrase; instead he maintained his description of a kite tail and flat, shiny, pie plates. Thus the term "flying saucer" seems to have been a headline writer's creative embellishment; yet subsequent reports in the press spread the term "saucers."

Over the course of the next month, there were hundreds of reports of flying saucers as well as many clear hoaxes. A July 4 sighting by a United Airlines flight crew was deemed to be particularly respectable and received more press coverage than Kenneth Arnold's initial story. The saucer sightings were quite varied, with some saucers said to literally be the size of pie pans and others the size of airplanes. While the original report mentioned a silver aircraft, subsequent ones were colorful and glowing.

Scientific speculation was extensive. A Los Angeles evening paper claimed that an unidentified physicist from California Institute of Technology had suggested that the saucers were experiments in the "transmutation of atomic energy." This was a reasonable (although scientifically uninformed) speculation for the time, coming as it did just a couple years after the public became aware of the power hidden in the nucleus of the atom. The atomic hypothesis was rejected by the chairman of the Atomic Energy Commission, David Lilienthal. He interrupted a reporter who was repeating the story and said,

"Of course, I can't prevent anyone from saying foolish things." A little later, Caltech issued a denial that anyone from that university had said flying saucers could be some sort of atomic experimentation.

Others speculated that the phenomena was mass hysteria or akin to sightings of the Loch Ness monster. Optical illusion, persistence of vision, and other similar causes were suggested.

One of the earliest suggestions of extraterrestrial origins came from a July 6 editorial in the *New York Times*, where the idea that "they may be visitants from another planet launched from spaceships anchored above the stratosphere" was airily dismissed. Arnold did say more than once that he considered that flying saucers could originate from somewhere other than Earth. On July 7, Arnold told the media that he had received lots of mail from people offering various explanations for what he had seen, from religious ideas to claims of extraterrestrial origins. In the *Chicago Times*, he is reported to have said, "Some think these things may be from another planet." He followed by noting that the speed at which the saucers maneuvered would induce acceleration forces that would kill humans. The story quotes further: "So, he too thinks they are controlled from elsewhere, regardless of whether it's from Mars, Venus, or our own planet." The ET idea had started to leak into the public arena. Arnold later mentioned the idea in a 1950 radio broadcast by Edward R. Murrow called *The Case for Flying Saucers*. He said, "If it's not made by our science or our Army Air Forces, I am inclined to believe it's of an extraterrestrial origin."

By July 7, 1947, saucer reports had come in from thirty-nine U.S. states and from overseas in Australia and many locations in Europe. However, the bulk of the observations came from the U.S. northwest. Across the nation, large and organized flights of pilots would take off from an area a hundred at a time to go out and look for unexplained aerial phenomena, with no success.

Things began to get silly, with clumps of dirt being reported as crashed saucers, or the top of a furnace, or saw blades with some electrical components welded to them. Some students soldered together two cymbals, tossed it in someone's yard, banged on the door, and ran away. An anxious inhabitant called the police and reported a crashed saucer. By July 18, the *New York Times* reported that the summer's chic women's hats were being modeled after flying saucers. And on July 8, a 25-year-old turtle named Flying Saucer won the eighth annual turtle derby in Chesterton, Indiana.

Eventually news coverage about flying saucers gave way to pieces that debunked some flying saucer sightings. For instance, there were examples of the

University of Chicago or Princeton University releasing high-altitude research balloons that were carrying instrumentation for meteorological or cosmic ray studies. Given the massive publicity about the saucers, it is inevitable that people would call the police and newspapers and report a new sighting. After a year or two, the media began to get a little bored with these reports, and they tapered off. But, by 1949, the military had concluded that there was something to the UFO reports and had agreed that they would put some resources into the question. This is unsurprising, considering how frequently the extraterrestrial idea made it into the papers. By far, the most common explanation for UFOs (other than imagination and hysteria) was secret weapons programs. The U.S. military knew that they weren't launching test planes that could fly more than a thousand miles an hour, so it is inevitable that those tasked with defending a country would want to find out if some other country had a new offensive capability that they needed to counter.

Of all the flying saucer reports that began in the summer of 1947, there was one special one that has lately managed to permeate the public's awareness more than any of the others. As you will see, the real story is somewhat different than what is commonly believed. So we turn our attention to a small town in southeast New Mexico: Roswell.

Roswell

The Roswell story is one of the best known in UFO-ology. In fact, if you'll forgive the bad joke, you'd have to be from Mars to not have heard of it. It goes something like this. A UFO crashed outside Roswell, New Mexico. Government agents, generally known as "the Men in Black," swooped into town and confiscated the flying saucer and the saucer's occupants, which included actual Aliens. The saucer and Aliens were transported to Area 51. One or more of the Aliens died and subsequently an autopsy was performed. A film of the autopsy was leaked in 1995 and shown on Fox TV. This most famous of Alien reports has been the obsession of UFO enthusiasts for more than 60 years.

There's only one problem. The fame of the Roswell incident is relatively new. It was forgotten for years. Here is what really happened.

It was the height of the Arnold-inspired UFO frenzy. On July 8, 1947, the Roswell *Daily Record*'s headline was "RAAF Captures Flying Saucer on Ranch in Roswell Region." The story describes how an unnamed rancher notified the local sheriff that he had an instrument on his premises. A major from the local Roswell Army Air Field took a detail of soldiers to the ranch and recovered the disk. After the local intelligence office inspected the instrument, it was flown to "higher headquarters" (quotes in original newspaper article). No details were released on the construction or appearance of the saucer.

The article went on to tell of another couple in town who thought they had seen a flying saucer. The saucer was supposed to be about 1,500 feet above the ground, travelling at 400 to 500 miles per hour and was estimated to be 15 to 20 feet in diameter. The man who observed the saucer was "one of the most respected and reliable citizens in town," and he kept the story to himself. According to the article, he had decided to tell people in town only minutes before word came around that the RAAF had a saucer in custody. It was very exciting stuff to have a flying saucer in hand.

The story changed the next day. On July 9, the Roswell *Daily Record* had a different headline "Gen. Ramey Empties Roswell Saucer," which is a cute way of saying "never mind." The paper ran two columns, first on the local sheriff, who was fielding dozens of phone calls from across the United States and Mexico, as well as three from England, including one from the London *Daily Mail*. The paper also identified the rancher by name. W. W. Brazel, who lived on the Foster ranch, was the person who found the remnants of the "so-called saucer."

Unfortunately for the UFO enthusiasts, the column also reported that the mysterious object found was a "harmless high-altitude weather balloon, not a flying disk." Even more specifically, what was found was a "bundle of tinfoil, broken wood beams and rubber remnants of a balloon." In the end, the UFO was identified as a specific type of weather balloon used to measure weather at altitudes much higher than the eye can see. The local army weather forecaster stated that they were identical to balloons he had sent up during the invasion of Okinawa to determine ballistic information for the heavy guns.

The news was not confined to Roswell. The Associated Press picked up on the story, and soon it was found in national papers. The *Chicago Tribune* reported more details on July 9 from the original press release, stating that "the many rumors regarding the flying disk became a reality yesterday when the intelligence office of the 509th [atomic] bomb group of the 8th air force, Roswell army air field, was fortunate enough to gain possession of a disk through cooperation of one of the local ranchers and the sheriff's office." However, it also reported that the mystery had been solved, when it identified the "flying saucer" as a "ray wind target."

The tone of flying saucer reports in the *New York Times* was rather skeptical throughout 1947, and those in Roswell were no different. On July 9, it did admit that the Roswell reports created more confusion than most, but it also told the official weather balloon story.

It is interesting that none of the reports I have read brought up the extra-terrestrial hypothesis. While there was massive confusion, the contemporary thinking seemed to be that the flying disks were unexplained phenomena or likely classified military objects.

Still, the report of a captured flying saucer must have energized the UFO community, correct? Especially a craft found so near the site of the first nuclear detonation? The report that the craft was spirited away, specifically to Wright Field in Ohio, surely got the attention of the alien aficionados, right? The conspiracy types must have had a field day? There's only one problem. That's not what happened.

Instead, the Roswell saucer simply disappeared from history. For 31 years, it was considered to be a false alarm; just a hysterical report in a hysterical time. And then 1978 rolled around.

In 1978, on what must have been a slow day in the newsroom, the *National Enquirer* simply reprinted the 1947 article from the Roswell *Daily Record*. UFO believers went nuts. Physicist and avid UFO-ologist Stanton Friedman tracked down the intelligence officer who retrieved the debris from the Foster Ranch and interviewed him. The intelligence officer's recollections made it into a 1979 documentary called *UFOs Are Real* and a 1980 *National Enquirer* article. This account didn't report a flying saucer, but it did talk about weird writing and flexible metal (which sounds a lot like aluminized mylar to a modern reader, although mylar was invented in the 1950s, long after Roswell).

That year also brought with it the publication of the book *The Roswell Incident*, which didn't provide much new information, mainly a lot of second-hand reports, supposition, and conjecture. It ended with the rather accurate statement:

Consider the implications of the Roswell Incident: If only *one* of the many individuals mentioned in this book who claimed to have witnessed the crash and/or subsequent recovery of an extraterrestrial vehicle is telling the truth—then perhaps at this very moment we sit at the verge of the greatest news story of the twentieth century, the first contact with live (or dead) extraterrestrials. This occurrence, if true, would be at least comparable to Columbus' encounter with the startled natives on his visit to the New World. Except for one thing. In this case we would be the startled natives.

The UFO community didn't forget Roswell, but everyone else did.

Things got really interesting in 1989 when the TV show *Unsolved Mysteries* devoted an episode that "reconstructed" what was supposed to have happened. This prompted a mortician from Roswell to contact Stanton Friedman and tell his story. The outcome of the subsequent interview was published in the 1991 book *UFO Crash at Roswell*, in which the now well-known story came into existence: alien bodies recovered, aliens walking around, small coffins, an army colonel making death threats, the disappearance of a nurse who knew too much, a dramatic series of events that make for an excellent story.

And, of course, there is the 1995 Alien autopsy as shown first in the United Kingdom and then on Fox TV. The show *Alien Autopsy: Fact or Fiction* purported to show an autopsy of a Roswell Alien. The show had twelve million viewers when broadcast in the United States. British entrepreneurs Ray Santilli and Gary Shoefield produced the show, and they claimed that they had purchased the film from a mysterious cameraman who had shot the original film in Roswell in 1947. However, in 2006, Santilli and Shoefield admitted in a documentary called *Eamonn Investigates: Alien Autopsy*, presented by Eamonn Holmes, that the film they had shown the world was not shot in 1947, but was rather as they called it, "a restoration." The claim is now that the original film had degraded beyond use, and they instead shot a new film, using fake alien bodies and a mix of animal parts. It is definitely a fake, as admitted by the producers, although they claim it is a faithful rendition of a true film that Santilli had seen earlier.

Now, in all fairness, many UFO enthusiasts have long believed the recording to have been a fake. But even if serious students of UFO-ology have dismissed the film, as have scientists, it doesn't change the fact that there are a lot of people out there who are simply casually interested in the subject and that this film has had an impact on the public. There are people who have only heard of the program who now wonder whether alien bodies are being held in Area 51 at Edwards Air Force Base in Nevada or in Hangar 18 at Wright Patterson Air Force base in Ohio.

The story has sufficiently penetrated the public's consciousness that what I've described here formed an important plot device in the 1996 film *Independence Day*, in which an Alien craft and bodies were being studied at Area 51, and was also the premise of the 2011 movie *Paul*, in which a laid-back, partying Alien escaped confinement. While he was confined, he had had a substantial impact on the science and technology of the second half of the twentieth century. Another nod to the Roswell history is a television show called *Roswell* (1999–2002), in which human teens and Aliens in the shape of human teenagers interacted. These are but a few examples of this story's working its way out into the public's awareness.

As one learns about Roswell, what is surprising is the fact that the story is a relatively new one. After lying dormant for about 30 years, it was revisited in the early 1980s and fell back into anonymity until the 1990s. It's really a fairly recent cultural phenomenon and one that the town of Roswell has eagerly embraced. If you go there, you'll be able to visit a museum devoted to the incident, with newspaper clippings on the wall, as well as assorted life-sized dioramas that depict various key scenes in the tale. You will find many shops devoted to selling alien-themed souvenirs. While not generally a fan of kitschy knickknacks, during my visit to Roswell, I admit to having been tempted to buy a bumper sticker that said, "Wear your seatbelt! It makes it harder for the aliens to suck you out of your car!"

Contacted!

George Adamski was what is sometimes called a "character." Maybe the best way to introduce him is with the opening of a book review of his 1955 book *Inside the Space Ships*, written by Jonathan Leonard in the *New York Times*. It begins:

> Competition is getting rough in the flying saucer business. Once a man could make an effect merely by seeing saucers. Then the saucers began to land. Now George Adamski has actually ridden in them. He was staying in a Los Angeles hotel when two men came to see him in a black Pontiac sedan. They looked like American business men and spoke English, but they were from Mars and Saturn (no antennae). They drove him to a softly glowing saucer in charge of a Venusian (no antennae). The saucer took off and flew on magnetic force to a mother ship 2,000 feet long that was hovering close overhead.

The review goes on with the same tongue-in-cheek style, describing the trip, the attractive Martian and Venusian women whom he met, and the philosophy they discussed. The trip, if true, sounds like an experience of a lifetime.

During the 1950s, Adamski gained notoriety in UFO circles and even to a degree in the public as the first of the "contactees" (i.e., people who claimed physical contact with aliens). In the twenty-first century, his claims are not considered reputable even among most of those who believe UFOs to be Alien visitors, but this wasn't always so. Adamski was a good-looking and charming man who had a fantastic story to tell.

Adamski was a self-described "wandering teacher." In the 1930s, he founded a school called the "Royal Order of Tibet," which taught self-mastery, utilizing a mixture of metaphysics and the occult. While he hadn't attended college, he was called "Professor" by his students and even signed some of his books with the title. When he moved to California, some of his students moved with him to continue to hear his teachings.

To appreciate Adamski, you need to read his books and to encounter his flamboyant writing. Let me sketch out the story told in *Flying Saucers Have Landed*, coauthored with Desmond Leslie and published in 1953. The story starts, "I am George Adamski, philosopher, student, teacher, saucer researcher." He claimed to have lived on Mount Palomar, home of the Hale Observatory, which housed the 200 inch telescope. He never worked there (in fact he was a handyman at a hamburger restaurant and lived 11 miles from the observatory), but people often associated "Professor" and "Palomar" and drew their own conclusions. In his book, he claimed that on October 9, 1946, he saw a gigantic spaceship hovering in the sky near his apartment. (Yes, that was prior to Arnold's 1947 sighting but recall that Adamski's book was published in 1953, and some of his stories could be characterized as creative.)

A couple of weeks later he was at work, telling people about what he saw, and six military officers were eating there. According to Adamski, the officers told him that his story wasn't fantastic and that, while they couldn't say anything, they knew the ship wasn't from this world. Although Adamski told several stories about seeing UFOs over the next few years, it was his tale of what occurred to him on November 20, 1952, that took his story much further than the typical UFO stories of the late 1940s.

It went like this: He was out in the desert with six companions looking for UFOs. He picked the location because he had a feeling of where to go, echoed by Richard Dreyfuss's character in the 1977 movie *Close Encounters of the Third Kind*. While standing in the desert, Adamski and his companions saw a gigantic cigar-shaped silvery ship, with orange all along the top. It drifted along and then went out of sight. According to Adamski, he told his companions, "The ship has come looking for me and I don't want to keep them waiting!"

He told them to wait for him for an hour and then set off into the desert alone. When he was far from the others, he saw a man standing at the entrance of a ravine, about a quarter mile away. Adamski walked toward him.

The person appeared to be an ordinary man, somewhat shorter than Adamski and with shoulder length, sandy-colored hair. His clothing was kind of like a high-necked jumpsuit, with elastic at the ankles and wrists. And he was very good-looking. Adamski reports, "The beauty of his form surpassed anything I had ever seen," and "in different clothing he could have easily passed for an unusually beautiful woman; yet he was definitely a man."

The person he encountered didn't speak English, but luckily Adamski was a believer in telepathy, for that was how they communicated. Through a combination of sign language and telepathy, he ascertained that the man was from Venus and that he was concerned with radiation coming out from Earth and that it might damage flying saucers. Adamski reasoned that cosmic rays in space were more powerful than those on Earth and, by reversing the logic, the radiation from the atomic bombs that were being tested on Earth was much amplified when it got into space.

Adamski then saw the flying saucer that brought the Venusian to Earth. Again through telepathy and sign language, he determined that the saucer was a scout ship and that the larger silvery ship he had seen earlier was the interplanetary craft. Adamski also asked the Venusian whether he believed in God. He did. Pretty heady discussions for two people with no shared language.

Further discussion revealed that all of the planets in the solar system were inhabited by humanoid Aliens and that the Aliens had taken people from Earth in their craft. Further, Venusians were immortal, although they could be killed. The immortality was of a form where their body died, but the spirit didn't and could move to another body.

After some additional discussion, the Alien made impressions in the sand by walking, leaving significant symbols on the ground. Luckily, Adamski had remembered to pack plaster of Paris in the car (you know . . . just in case), and he and his companions later made casts of the symbols.

Adamski was made to understand that he couldn't go in the ship, so he walked the Alien back to the saucer. Adamski had taken some pictures that were affected by some sort of emanations from the scout ship's propulsion system (which explained why the developed pictures were of such low quality). Prior to leaving, the Venusian took some of Adamski's film and somehow made Adamski understand that it would be eventually returned. The Alien entered the saucer and left. Adamski never did get the Alien's name, and he returned to his companions.

Later, he told everyone of his experience, including reporters. The book claims his story was carried in the November 24 edition of the *Phoenix Gazette*. (This part of the story is true, although the article began in a rather tongue-in-cheek style and had significant differences in the details as reported by Adamski in his book, for instance there is no mention of telepathy, rather the Alien spoke a mixture of English and a language that sounded like Chi-

FIGURE 2.2. An artist's depiction of the saucer George Adamski claimed to have photographed. This saucer is reported to be the craft piloted by Aliens who later became known as the Space Brothers.

nese.) He later had his film developed, and the photos showed the saucer, which looked like some sort of light fixture with three light bulbs under it (figure 2.2).

A few weeks later, Adamski said he was home when an iridescent glass-like craft flashing with brilliant colors moved through the sky toward his house. Apparently the Aliens knew where he lived. When the ship moved over him at a height of about 100 feet, a porthole was opened and a hand dropped the film. The ship left. When Adamski had the film developed, it was covered with symbols that were "still being deciphered."

Here Adamski's first tale ended. He says that the Aliens are friendly and that they mean "to ensure the safety and balance of the other planets in our system." However, "if we continue on the path of hostility between nations of Earth, and if we continue to show an attitude of indifference, ridicule and even aggression toward our fellow-men in space, I am firmly convinced that they could take powerful action against us, not with weapons of any kind, but by manipulation of the natural force of the universe which they understand and know how to use."

A cautionary tale, indeed. The similarity to the closing speech of *The Day the Earth Stood Still* might be a coincidence. Or not.

The story told in *The Flying Saucers Have Landed* took only a couple of dozen pages. However Adamski's 1955 book *Inside the Space Ships*, was far more adventurous. It tells that he encountered Aliens dressed as businessmen at a hotel in Los Angeles, who drove him around in a black Pontiac sedan. He also flew in space with: a Venusian named Orthon, a Martian named Firkon, and a Saturnian called Ramu, three members of a civilization called "Space Brothers," named for their harmonious culture. He assures us that these aren't their real names, but how he decided to name them. On the ship, he met the two previously mentioned lovely women from Venus and Mars. He also goes on to describe their deity, about which he and the Aliens had long philosophical and religious discussions. It is unsurprising perhaps that these "Space Brothers" told of a cosmic brotherhood and affirmed that Adamski's teachings of the 1930s were exactly correct. This incredible similarity is the dominant reason why so many people are suspicious of his story. Well, that and the two beautiful and statuesque female "Venusian bodyguards" that accompanied him on his lecture tours.

For tour Adamski did, as did his coauthor Desmond Leslie. Perhaps the most notorious engagement was in May 1959, when Adamski had a private audience with Queen Juliana of the Netherlands. Juliana had a reputation for being interested in faith healing and similar sorts of sketchy phenomena.

Adamski's popularity waned in the 1960s when the Soviet space probe *Luna 3* showed barren wasteland on the far side of the moon, where he had reported snow-covered mountains. His response? The Soviets had faked the photos. More than one wag has claimed he should recognize a faked photo when he saw one. We now realize that Venus is quite the antithesis of the paradise that he claimed in his books, although modern Adamski believers have noted that he did state that the Venusian cities were underground and at least a few diehards have invoked parallel dimensions and claimed that Orthon's home isn't in our universe.

It's easy to see the allure of Adamski's message. His Aliens are recognizably angelic (albeit without the halos). The Space Brothers believed in peace and harmony and hoped that humanity would eventually join them in cosmic brotherhood. Adamski's antinuclear message also resonated with an American public who well remembered the destruction of World War II and were quite worried about the territorial ambitions of the nuclear-armed Soviet Union. Given his poor track record at predicting the environment on other planets in the solar system, Adamski is now considered an unreliable prophet, but his message of peace and cosmic harmony did spawn imitators and some of them exist to this day. For instance the Raëlians follow the teachings of the

French journalist Claude Vorilhon, who now calls himself the prophet Raël after a purported 1973 encounter with an Alien called Yaweh.

Adamski died on April 12, 1965, but it appears that death was but an ephemeral state for him. As reported in Eileen Buckle's 1967 book *The Scoriton Mystery*, another contactee by the name of Ernest Bryant claimed to have met three Space Brothers on April 24, 1965. One of the three Aliens he met was a young man named "Yamski," who was supposed to be George Adamski, returned to bodily form.

One can abbreviate Adamski's life story the following way: A charismatic man of modest origins claimed to have discovered a way to live a fulfilling and enlightened life. He gathered acolytes around him whom he taught. One day, he took a few of his students and a couple of people interested in his teachings into the desert to a location that he knew intuitively to be right. He separated from his companions and ventured alone into the desert where he met an angelic being who told him cosmic truths and left him with a cryptic message in the earth, which Adamski brought back for interpretation. After a life of speaking to larger groups, bringing them the message of peace from the angelic beings from above, Adamski died, only to rise from the dead twelve days later in a different form to speak to a true believer.

Stated that way, it is rather unsurprising that some religions or cults have sprung up around Adamski and his teachings, no? Even if not directly related, other groups have begun with a similar message, including the Aetherius Society and Raëlism, although there are others. The teachings of the Aetherius Society blend earthly religions, yoga, "spiritual batteries" (which can avert disaster), and an extraterrestrial messiah who will one day bring humanity into the community of the stars. Some, like the Heaven's Gate cult, espouse not quite the same message as Adamski's teachings but do incorporate extraterrestrial elements. Scientology claims that 75 million years ago, a leader named Xenu destroyed billions of Aliens in atomic explosions here on Earth, and those souls, called thetans, are among us. These quasi-religious beliefs have had modest impact on society's vision of Aliens, but they pale in comparison to our next story.

Abducted!

If contact by space angels is an uplifting and spiritual experience, not all interactions with Aliens are as positive. The next paradigm in the saga of Aliens on Earth began in 1961, when Betty and Barney Hill encountered a new brand of extraterrestrials.

An interracial couple in conservative New Hampshire in 1961, they were a little more liberal than some of their neighbors. But otherwise the Hills were pretty ordinary. Betty was a social worker, while Barney worked for the postal service. They were active members in a Unitarian Universalist congregation and the NAACP. Sometimes nicknamed "Ma and Pa front porch," they inspire more credibility than do some of the attention-seeking people we encounter in the UFO saga.

Betty and Barney were driving south through New Hampshire on their way home from vacation. It was about ten o'clock on the night of September 19, 1961, and they had just stopped for dinner in Colebrook. A cup of coffee and a cigarette for alertness and then they were back on the road, Barney behind the wheel.

They drove on, expecting to get home around three o'clock in the morning. Betty noticed a bright star or planet near the moon, not at all an unusual occurrence. When she looked at the moon some time later, she saw a second star near the first one. However the second star seemed to be getting bigger. They dismissed it as probably a satellite or something.

They had brought their dog along for this trip, and she began to get antsy. So Betty wanted to take her out for a walk. They parked at a spot where there was a good view of the sky and grabbed some binoculars. The object was moving all right, but what it was, was hard to know.

They got back in the car and continued to drive. Betty kept an eye on the light in the sky, while Barney watched the road. Betty realized that the light couldn't possibly be a satellite as its path was erratic. Barney dismissed it as a plane, but the fact that the light was following them made that seem less and less likely. In addition, as the light got close, there was no sound of a plane's engine.

Things started to get a little weird. The light came to within a couple hundred feet of the car. Betty put the binoculars to her eyes and was shocked to see what appeared to be windows along the sides of the light. The light was no longer a light, but some sort of craft, with complex structures and "shaped like a pancake." She made Barney stop the car, and he took a look at it himself.

While Betty stayed with the car, Barney set off across a field with the binoculars to get closer to the aircraft. He saw half a dozen figures inside the craft watching him out the portholes. They were wearing uniforms and braced themselves against the windows as the saucer tipped toward them. Most of the figures turned to what appeared to be a control panel, while one—presumably the leader—continued to watch him.

Barney panicked and ran back to the car, put it in gear, and headed off down the road. He told Betty to keep an eye on the saucer, but she couldn't see it anymore. He thought that the craft might be above them. Then, suddenly they heard a beeping sound coming from the vicinity of their trunk. They didn't know what was causing it. But suddenly things got fuzzy and they both got drowsy.

After a time—and they weren't sure how long that was—they heard the beeping again. As they came out of their mental fog, they found themselves driving along the road. Still a bit groggy, they entered Route 93 and saw a roadside sign that said "Concord 17 miles." They had gone about 35 miles.

They talked a little bit as they drove, and Betty asked him if now he believed in flying saucers. This was a question she had asked in the past when a story had appeared in the newspapers. Barney, ever the skeptic, told her not to be ridiculous. When they eventually got home, Betty read the clock in their house, which said it was a little after five in the morning, some two hours after their expected arrival time.

That's the first part of the story. The rest is much weirder. However, this we will sketch in less detail. The reason is that the next chapter in the story developed slowly and after interaction with many people. The interested reader should read John Fuller's *The Interrupted Journey* or Stanton Friedman and Kathleen Marden's *Captured: The Betty and Barney Hill UFO Experience*.

The Hills couldn't account for the two hour gap in time. Betty talked to her sister, who had earlier reported a UFO encounter. Her sister spoke to a police captain, who suggested that Betty speak with the air force. Barney wanted nothing to do with it, but Betty called Pease Air Force Base and made a report. The next day, they were called back by the reporting officer who wanted to confirm some details. Barney started to warm up to the idea of talking to people about it.

Betty's curiosity about UFOs was stronger than before. She went to the library to find whatever she could, including *The Flying Saucer Conspiracy* by Major Donald Keyhoe. The thesis of the book was that either the UFO phenomenon was a mass hysteria, well worth study, or that it was real, which was more interesting still. Keyhoe found the second thesis more plausible and was convinced that the air force was covering up the many reports that they had received. Keyhoe's book told many stories, including an abduction one (which will be relevant soon). Betty was interested enough that she wrote Keyhoe asking whether he had other writings she could read. Betty was hooked.

With that fateful letter, Betty made herself known to the UFO world. Keyhoe passed along the letter to a UFO researcher at the Hayden Planetarium. The researcher spoke to the Hills and wrote a report, which was submitted to the National Investigations Committee on Aerial Phenomena (NICAP), an organization started by Keyhoe to investigate UFOs. Through these connections, the Hill story started becoming familiar to the community of UFO aficionados.

It should be noted that the Hills were not media hounds. They didn't talk to reporters. They spoke to governmental agencies and UFO investigators of the disciplined variety. The Hills wanted to know what happened to them. And, across the board, the really pressing question was "where did those two hours go?"

About ten days or so after the incident, Betty started having vivid dreams that when she and Barney were outside the car, they were escorted into the saucer, where they underwent medical tests, including the insertion of a needle into her navel to test for pregnancy. The examiners were short, between 5′ and 5′4″ tall. They were gray with bluish lips and huge noses "like Jimmy Durante." They were very human in their appearance and dressed in military-style uniforms, with hats like those worn by the U.S. Air Force. She wrote the dreams down in November 1961.

Quite aside from questions of UFOs, Barney was stressed. His job was on the south side of Boston, with a 120-mile daily commute, round trip. He worked the night shift and was unable to spend much time with his sons from his first marriage. In an effort to better cope with his stresses, Barney went into therapy. In late 1963, Air Force Captain Ben Swett, who the Hills met at a presentation at their church, suggested that they ask the therapist if hypnosis would help. The therapist referred the Hills to Dr. Benjamin Simon. When Simon spoke to Barney, it became obvious that the saucer encounter was causing Barney more problems than he would admit, so Simon decided to hypnotize Hill to perhaps understand what happened in the two hour gap. The hypnosis went on over a period of 11 months.

Simon hypnotized both Barney and Betty separately to avoid contaminating their recollections. Barney went first. Under hypnosis, he remembered an encounter much like Betty's dreams. This was two years after the incident, and Barney had no doubt spoken to Betty at length about them, although there were differences in their accounts. Barney remembered the Aliens (for by this time it became clear that Aliens were behind all of this) were short and gray, but without a nose. They spoke to him in English but without moving their mouths. Barney called it "thought transference," as he was unfamiliar with the term telepathy. Betty and Barney were examined in different rooms in the saucer. During the examination, the Aliens studied the Hills' physiology, spending a large amount of time on the pelvic region, including putting a cup of some kind over his genitals to extract a sperm sample and inserting some sort of tube in his anus. Eventually, he was returned to Betty and they both returned to their car, in the manner of sleepwalkers.

Betty recalled similar details to Barney's during her session. Under hypnosis, both of the Hills' accounts were closer to each other's than they were to Betty's dreams as written down previously. After the examination, Betty asked the Alien where they came from, and he produced a star map. Simon was able to implant a posthypnotic suggestion for her to draw the map, and she did so. Simon also had Barney draw a picture of an Alien while under hypnosis. The hypnosis sessions ended in the summer of 1964, although the Hills and Simon kept in occasional contact through 1965.

Simon's conclusion was that the recollections were simply a repeat of Betty's dreams. He did not believe that they had been abducted by aliens. He wrote up the account in the journal *Psychiatric Opinion* and the Hills went about their normal lives, feeling much better now that they felt that they could account for the missing time. The Hills would still talk about their experience with friends and family and the occasional UFO researcher, but they didn't seek out the media. To this point, the Hill episode was only a curiosity discussed by UFO enthusiasts. This was about to change.

Reporter John Lutrell of the newspaper *The Boston Traveler* had heard of the Hills and obtained a 1963 audio recording of them talking about their experience. He did a little digging and found that they had spoken to Simon and asked for information. Simon and the Hills refused to cooperate, so Lutrell reported with what he had available. On October 25, 1965, he published the paper *UFO Chiller: Did THEY Seize Couple,"* the first of a three-part series. UPI picked up the story on the following day, and the Hills became international celebrities.

The Hills were aghast at the report and decided to tell their story. Writer John Fuller worked with them in 1966. The result was the highly successful *The Interrupted Journey: Two Lost Hours aboard a Flying Saucer.* The book contained some sketches Betty had made of the star map and others that Barney had drawn showing what their captors looked like. Later critics compared the Hill's account of the appearance of the Aliens with those in an episode of the television show *Outer Limits,* broadcast just a few days before the relevant hypnotic session (figure 2.3).

FIGURE 2.3. Barney Hill's drawing of the alien he believed he saw (*left*) is the progenitor of the public's modern concept of Aliens. The middle figure appeared in "The Bellero Shield," an episode of the television show *Outer Limits*, and is thought by some to be the inspiration for Barney's drawing. The figure on the right is from the 2011 movie *Paul* and shows a modern depiction of a typical Alien. Copyright John G. Fuller (*left*), United Artists Television (*middle*), Universal Pictures (*right*).

In 1968, amateur astronomer Marjorie Fish read *The Interrupted Journey* and was interested in the star map. Over a period of 5 years, she made a three-dimensional model of stars near Earth, using beads and string. She even visited Betty Hill in the summer of 1969 (Barney having died earlier that year) to get as much information as possible. When the model was complete, she walked around the model with Betty's map in her hand. She finally found an angle that seemed to match. She concluded that the Aliens had come from Zeta Reticuli, specifically Zeta Reticuli 1, as it is a binary star system.

This hypothesis reached the editor of *Astronomy Magazine* and, for the first time, this magazine published a UFO story in December 1974. It compared contemporary astronomical knowledge, including all sunlike stars within a sphere centered on the Solar System with a 55 light-year radius, to Fish's map. The article concluded that the reconstruction was pretty good. Companion articles discussed the metallicity of the stars in the Hill map as identified by Fish. Zeta Reticuli 1 and 2 are deficient in metals (60% that of the sun, using the astronomer's definition of metals as "everything that isn't hydrogen and helium"). This doesn't rule out these stars as a host for a technologically advanced species, but it does make it harder. After all, you need metal to make flying saucers and other elements to make the Aliens themselves. In addition, over the course of the following year, there was active discussion in the letters to the editor column, including contributions by Carl Sagan and his research associate Steve Soter.

Another way the public heard the Hills' story was the 1975 made for TV movie called *The UFO Incident*. The dramatization was an approximately faithful depiction of *The Interrupted Journey*. Whether or not the Hills encountered Aliens that night, their story is the archetype for Alien abduction stories: the amnesia, the examination, the fascination with the human pelvic region, the small gray humanoids, the big black eyes. In short, Betty and Barney Hill told us what Aliens look like.

Ancient Aliens

Carl Sagan is not a name that one normally associates with terrible science, but it is possible that he had an unintentional hand in launching a surge of books that advanced the theory that not only has the Earth been visited by Aliens but that these visits began thousands of years ago. In a 1966 book *Intelligent Life in the Universe*, astrophysicists Carl Sagan and Iosif Schklovsky included a chapter devoted to urging the archaeological community to be open to the idea that the Earth had been visited by ancient astronauts in the past. They didn't claim that it had happened, but simply that it was a possibility to be considered. Other authors weren't as cautious in their claims.

Erich von Däniken is a Swiss author who has the distinction of being the person who blasted the idea of ancient astronauts into the public consciousness. His 1968 book *Chariots of the Gods* was a smash success, with some 20 million books sold to date, and he has published nearly twenty books in English. He has also been jailed three times for fraud and theft. A criminal record is not a reason to a priori dismiss a person's ideas but, given the outlandish nature of von Däniken's claims, a record that includes fraud is presumably a relevant bit of information.

The central thesis of his books is that there is tremendous evidence for Alien visitation in the archaeological and historical record. He suggested that in the Christian Bible, the chariot of Ezekiel was a report of a UFO as seen by Bronze Age eyes. He interpreted the lid of a sarcophagus of a Mayan king as depicting an astronaut piloting his craft. The Great Pyramids of Giza, the Nazca Lines in Peru, Stonehenge, the huge heads on Easter Island— there aren't many interesting large ancient monuments that have eluded his speculation.

Few, if any, archaeologists give any credibility to von Däniken's theories. Most of his claims have been debunked, and von Däniken himself has conceded in interviews and in documentaries that some of his claims were false, embellished, or since discredited. Here are some examples. A picture in *Chariots of the Gods* is claimed to be reminiscent of a runway and parking areas for spaceships. Closer inspection shows that the picture in the book was cropped in a quite misleading way and that the parking areas were just too small to park much of anything, with the "runway" being 8 to 10 feet across and the "parking lot" being not much bigger. In his book *Gold of the Gods*, he tells of an expedition in which he was guided through tunnels containing gold, statues, and a library in a cave in Ecuador.

In an interview in the December 1974 issue of *Playboy* and again in the 1978 *Nova* episode "The Case of the Ancient Astronauts," he admits to not actually having been in the cave and having embellished his story to make it more interesting. In the same documentary, he defends a museum in which carvings reported to be thousands of years old are stored. The documentary's producer located a local sculptor who claimed to have made the carvings and who re-created some of them for the camera. It should be stressed that von Däniken did not participate in this fraud, which seems to have been the work of a local entrepreneur out to make a buck, but von Däniken clearly isn't one for letting something as inconvenient as the truth get in the way of a good story.

The *Playboy* interview should be required reading for von Däniken enthusiasts, as it clearly demonstrates a shockingly cavalier attitude toward disciplined investigation. Even though von Däniken has conceded that many of his claims in his books didn't stand up to even casual scrutiny, later versions of his books remain uncorrected. It would seem that diligent scholarship isn't an important consideration for these publications.

Regardless of the veracity of his claims, there is no question that von Däniken's books had a huge impact on the public. This impact was amplified by a subsequent German language film version of his book *Chariots of the Gods*. This film was subsequently edited, dubbed into English, and shown in 1973 on American television under the name *In Search of Ancient Astronauts* with *Twilight Zone's* Rod Serling doing the narration.

Von Däniken is not the only author to postulate ancient Aliens. In his 1976 book *The Sirius Mystery*, Robert Temple tells of the Dogon tribe in Mali who are reported to have long believed that the star Sirius has a companion that orbits the main star with a period of 50 years. Western astronomy discovered a faint companion star in 1862 that is invisible to the naked eye. As it happens, this star has an orbital period of about 50 years. Temple took this interesting bit of information and added claims about the origins of the culture of ancient Egypt and Greece, to name but a few. Temple didn't say he was certain that ancient astronauts gave the Dogon their knowledge, as an earlier, undiscovered, human culture with advanced technology could also explain the mystery. Temple did say that he thought that the Alien hypothesis was the more likely of the two.

Naturally, some anthropologists criticize the ethnographical studies on which Temple based his book, stating that the Dogon did not have a multicentury fascination with Sirius. Others claim that the origins of the knowledge of Sirius B stems from cross-cultural pollination, specifically from Europe (and possibly from the original ethnographers). Temple's book didn't penetrate the public consciousness the way von Däniken's book did, and so we leave it without deeper discussion.

The idea of ancient astronauts has certainly entered the public awareness. This can be seen in the 1994 movie *Stargate*, in which the ancient Egyptian civilization was modeled on Alien visitors to Earth millennia ago. The movie spawned three television shows with more than 250 episodes, spanning 14 years. We will revisit this series in chapter 4.

Aliens Today

In this chapter, we have taken a whirlwind tour through what we might call "Alien-ology." The incidents here are by no means the only tales of Alien contact there have been, nor are they the first. The stories here are not even selected as being ones that are plausible. They were selected as the stories that grabbed the public's attention and shaped our collective vision.

There are still people who believe in all of the tales told here and in many others. In the next two chapters, we'll tell the story of Aliens in fiction and relate the fiction to these tales of supposedly true extraterrestrial contact. But perhaps it is worthwhile to list the most common forms of Aliens that one will encounter if one attends a UFO convention. (Note that we will repeat this exercise at the end of the next chapter to include Aliens that come mostly from science fiction.) The typical aliens are:

Little green men. These are not terribly common anymore and originated more in the fiction of the early twentieth century. LGM were diminutive humanoids, sometimes with antennae. They were the precursors of the Grays.

Grays. Grays are the aliens of Betty and Barney Hill. They are short humanoids, gray in color, with large heads, no noses, pointy chins, and large, almond-shaped, soul-less black eyes. (Betty's dream of large-nosed aliens morphed over time into our now-familiar Grays.) They abduct humans and perform medical tests on them, frequently centered around the pelvic region.

Nordics. Also called Space Brothers, these Aliens are bigger than humans, beautiful in countenance, and spiritual in nature. They contact humanity only to teach us of the harmonious ways of the peaceful space community. These are Adamski's Aliens, although in Adamski's original contact, the Space Brothers weren't larger than humans.

Reptilians. These are a lesser-known form of Aliens, so they don't warrant a special segment. They tend to be much larger than humans (5 to 12 feet tall); they drink blood and can shape shift. According to British writer David Icke, they live on Earth in underground bases and have created reptilian/human hybrids. Most of the world's leaders are hybrids, including ex-U.S. president George W. Bush and Britain's Queen Elizabeth. The origins of this alien stem from a 1967 abduction report in which the Aliens had a slightly reptilian appearance and had a winged-serpent insignia on their uniforms.

Wrap Up

Throughout this chapter, I have described the incidents over the past 60 years or so that have shaped what we, as a culture, think about Aliens. I have not troubled myself to be skeptical, although I personally am unconvinced by any of them. For our purposes, the question of whether these encounters are real, intentional hoaxes, or well-meaning mistakes is quite unimportant. What is important is that these episodes are the ones that have defined society's vision of Aliens.

Skeptics will point to various things, for instance the episode of *The Outer Limits,* which portrayed Aliens looking much like Betty and Barney Hill reported and played just 12 days before the day they were hypnotized and described Aliens with big eyes and no nose to the therapist. These were features not present in Betty's dreams. Skeptics will also point to the fact that Kenneth Arnold didn't call the phenomenon he saw a "flying saucer." That fact was misinterpreted by a headline writer, and yet subsequent sightings were saucers and not the shape that Arnold saw. And, of course, there is Adamski's usurpation of the classic prophet story and von Däniken's astoundingly cavalier attention to archaeology. There are many books and countless articles out there deconstructing the Alien tale and it is entirely fitting if you are skeptical.

But it doesn't matter. These are the people and tales who have told us all what Aliens look like.

THREE

FICTIONS

Write me a creature that thinks as well as a man, or better than a man, but not like a man.

John W. Campbell, Editor of *Astounding*

Literature is one of the many fine creations of humanity. In writing fictional tales, an author can take us to places we've never been before or show us a situation that perhaps we've never considered. A good story can show familiar themes in unfamiliar surroundings and, done well, a story can be a lovely metaphor, a story in which the real message is unspoken, but stated clearly even so.

Of all the types of literature that have been developed over the millennia, science fiction is unique. It allows for plot devices that are not available to other genres. Its only possible competition is fantasy, but even that type of tale has at least some strictures. In science fiction, almost anything is allowed, as it can incorporate all of the other forms of literature with a setting that is quite unreal, say a murder mystery involving an alien death ray on Betelgeuse or a pair of star-crossed lovers who were literally born under different stars.

As we move forward, it bears remembering the theme of this book, which is the evolution of mankind's vision of Aliens. Thus the subthemes of science fiction that describe the impact of robotics, future dystopias, and even space travel into an empty galaxy (for instance Isaac Asimov's epic *Foundation* series) are not really topics suitable for our discussion. Further, it must be noted the stories that are considered the most important by the more serious science fiction fans (e.g., *Foundation* or Frank Herbert's *Dune* or much of Robert Heinlein's Lazarus Long stories) are not always the ones that have had the biggest impact on the public. The stories that influence public thinking are the ones that are disseminated most widely, which generally means radio, television, or movies. Delightful and innovative stories that live only in pulp magazines and science fiction anthologies are often read by a small group of people.

Accordingly, in this chapter and the one that follows, we will focus on the high-impact, high-visibility stories and try to understand what made them popular. This is in no way a simple job. Science and science fiction interact with one another in a way that cannot be easily disentangled. In a similar way, a popular movie may beget other movies that are clearly derivative. This can feed back into the science fiction literature, until it is difficult to know how the tale actually started.

We begin at the turn of the twentieth century. As we travel through the decades, trying to understand how our modern thinking of Aliens has developed, we will consider books, pulp periodicals, radio shows, movie short serials, and feature films and television series. We will see that the dispersal of ideas about Aliens is clearly interwoven with the existence and growth of mass media. Just as the moon hoax of 1835 needed the penny press to become a widespread phenomenon, so too have our modern visions depended heavily on the growth of visual media, especially television and movies.

Though he was not the first to combine scientific knowledge—or theory—and storytelling, perhaps Jules Verne might still be called the real father of science fiction. His stories were published in the 1870s and include such famous titles as *Twenty Thousand Leagues under the Sea* and *A Journey to the Center of the Earth*. However his *From the Earth to the Moon*, with the sequel *Around the Moon*, are the only stories that deal with extraterrestrial travel. In the story, a crew is shot from a big cannon, they orbit the moon and return to Earth. But no Aliens were encountered in Verne's stories, so we turn our attention elsewhere.

Mars Attacks the First Time

Arguably, the father of Alien fiction might be H. G. Wells. Wells was trained as a science teacher and was an editor of the science journal *Nature* during the

Martian canal frenzy. His 1898 story *The War of the Worlds* reflects speculation of the sorts championed by Percival Lowell and tells of an invasion of Earth by Martians against which mankind has no defense. It is a testament of the quality and timeliness of the story that it has never gone out of print. The story begins with a dying Mars. Martians, being the older and more technologically advanced civilization, shoot cylinders toward Earth by means of a large cannon. The cylinders land in England and, after a brief excursion outside their spaceship, the Martians return to their craft, only to emerge later in large tripods, small craft perched in the air on three long legs "higher than many houses." The craft also sported articulated tentacles that could grasp things and a heat ray that could disintegrate what it touched. After a long and harrowing tale of death and destruction, the Martians eventually died, laid low by a disease from Earth. As the Martians had no immunity to illness, mankind's salvation was but luck.

While most of the story spoke of the battle between the tripods of the technologically advanced Martians and the hapless humans, we were briefly introduced to the Martians themselves. The narrator expected to see a humanoid, "everyone expected to see a man emerge—possibly something a little unlike us terrestrial men, but in all essentials a man." However, the Martian was in fact rather unlike humans. The Martian was a big, grayish, rounded, bulk, the size of a bear and covered with tentacles (figure 3.1). It had two large and dark-colored eyes, with a mouth that panted and dropped saliva. The narrator relates:

> Those who have never seen a living Martian can scarcely imagine the strange horror of its appearance. The peculiar V-shaped mouth with its pointed upper lip, the absence of brow ridges, the absence of a chin beneath the wedgelike lower lip, the incessant quivering of this mouth, the Gorgon groups of tentacles, the tumultuous breathing of the lungs in a strange atmosphere, the evident heaviness and painfulness of movement due to the greater gravitational energy of the earth— above all, the extraordinary intensity of the immense eyes—were at once vital, intense, inhuman, crippled and monstrous. There was something fungoid in the oily brown skin, something in the clumsy deliberation of the tedious movements unspeakably nasty. Even at this first encounter, this first glimpse, I was overcome with disgust and dread.

Wells's story was published in a serial format in *Pearson's Magazine* in 1897, followed by the appearance of the book in 1898. As was typical of the time, a serialized novel appeared over the course of several issues, with each install-

FIGURE 3.1. These illustrations by Alvin Corréa from the 1906 edition of H. G. Wells's *War of the Worlds* show early depictions of Aliens and their spacecraft. Martians were octopus-like in appearance, while their fighting machines maintained some of the fluid movements that would appear natural in an invertebrate. The Martians were handicapped by being under the much larger Earth gravity.

ment ending in a cliffhanger of some sort so as to induce the audience to buy the next issue. The story resonated well with readers who were beginning to worry about the *fin de siècle* (French for "end of the century," the equivalent of the late twentieth-century Y2K worries). The thinking was that the culture was in decline and awaiting an invigorating renaissance.

The War of the Worlds is noteworthy not only for its impact in 1898 England but also for its other intrusions on popular culture. On Halloween 1938, 23-year-old George Orson Welles was a rising young director and producer. He was also to unleash what may be the most famous radio broadcast of all times. In an era before television, families crowded around the radio and listened to news, music, and entertainment. On the CBS radio show *The Mercury Theatre on the Air*, Welles broadcast the now-famous version of *War of the Worlds* to the radio audience. At the beginning of the show, the narrator said that the story was set in 1939 (i.e., a year in the future), but not everyone heard that. Being broadcast at the same time was the rival (and more popular) Edgar Bergen / Charlie McCarthy program. However, radio channel surfing was just as popular in the 1930s as television channel surfing is today. Some people would cut away from the Bergen/McCarthy show to see what was going on a different channel. And they tuned into what seemed to be a news broadcast saying that Martians had landed in Grover's Mill, New Jersey, and were attacking. The secretary of the interior was quoted as saying:

> Citizens of the nation: I shall not try to conceal the gravity of the situation that confronts the country, nor the concern of your government in protecting the lives and property of its people. However, I wish to impress upon you—private citizens and public officials, all of you—the urgent need of calm and resourceful action. Fortunately, this formidable enemy is still confined to a comparatively small area, and we may place our faith in the military forces to keep them there. In the meantime placing our faith in God we must continue the performance of our duties each and every one of us, so that we may confront this destructive adversary with a nation united, courageous, and consecrated to the preservation of human supremacy on this earth. I thank you.

Scary stuff.

Welles did try to counter potential panic, for he ended the broadcast with:

> This is Orson Welles, ladies and gentlemen, out of character to assure you that *The War of The Worlds* has no further significance than as the holiday offering it was intended to be: the Mercury Theatre's own radio version of dressing up in a sheet and jumping out of a bush and saying "Boo!" Starting now, we couldn't soap all your windows and steal all your garden gates by tomorrow night, so we did the best next thing. We annihilated the world before your very ears, and utterly destroyed the C. B. S. You will be relieved, I hope, to learn that we didn't mean it, and that both institutions are still open for business. So goodbye everybody, and remember, please, for the next day or so the terrible lesson you learned tonight: that grinning, glowing, globular invader of your living room is an inhabitant of the pumpkin patch, and if your doorbell rings and nobody's there, that was no Martian—it's Halloween.

The radio show was short, only 60 minutes, but the newscast style gave verisimilitude to the tale. People believed it and there was some panic, although there is some modern debate as to the degree of the actual alarm. There were some twelve thousand newspaper articles over the next month on the impact of the broadcast. It could be that the country was primed for tales of battle and destruction. A front page article in the October 31 late edition of the *New York Times* was headlined *Radio Listeners in Panic, Taking War Drama as Fact*. As a contrast, the article immediately to the right of it was titled *Ousted Jews Find Refuge in Poland after Border Stay*. Hitler and the Nazis were beginning their move. The Anschluss, which annexed Austria to Germany, had occurred in March 1938. The occupation of the Sudetenland in what was then Czechoslovakia had taken place in early October 1938, after Western powers abandoned the country. The war drums were beating, and a full-scale invasion of the United States didn't seem as ridiculous as it might today.

The War of the Worlds was also dramatized in a 1953 movie, in which the Martians had landed in southern California, and their technology had evolved to be able to resist an atomic detonation. The story is similar to the original book, and the Martians again died due to fatal susceptibility to Earth microbes. It was the most economically successful science fiction movie of the year and won three Academy Awards, including one for special effects. The 1953 version appeared right after the UFO frenzy of the late 1940s and during a period of popularity of movies involving flying saucers, space, and Aliens. The movie provided fodder for multiple screenplays, most recently Steven Spielberg's successful 2005 version.

Into its second century, *The War of the Worlds* persists in popularity less for its depiction of Martians than for its drama and its portrayal of mankind's response to adversity. The story is timeless, but it has surfaced to popular approval at times when it resonated with the public.

Barsoom

If *The War of the Worlds* told us about Aliens, it told us but little. For most of the book, the invaders were faceless enemies, ensconced inside their craft as they ravaged the countryside. The walking tripods, with their powerful ray guns, can be seen as a metaphor for the later Panzers and Stukas of the Nazi blitzkrieg or the more recent "shock and awe" of American escapades in Iraq. Faceless, mechanized, enemies went where they wanted with near impunity. The tripods could have been replaced by robots.

For a different vision of Martians, we need to turn to another author, Edgar Rice Burroughs. Burroughs was born in Chicago and spent some time in the U.S. army in Arizona. After a medical discharge, he drifted for some years, doing menial jobs. The year 1911 found him working as a pencil sharpener salesman. It was then that he began to write. His first story was *Under the Moons of Mars* and was serialized in *The All-Story*, a monthly pulp magazine. This story was retitled *A Princess of Mars* when it appeared several years later in book form. As his Mars story was coming out, Burroughs also wrote the first of the Tarzan series, published in the same periodical. Burroughs eventually wrote about seventy books and pioneered the idea of exposing his stories across many media outlets, from books, to serializations, to comics and movies. The public couldn't get enough of Tarzan, and it's a story we still know today.

However it was in *A Princess of Mars* that he wrote of life on our sister planet. He eventually was credited with eleven books in the Barsoom series, with some others written by his son. While these books were never intended to be taken seriously, they were written as if they were totally true stories. The hero, John Carter, was introduced as Burroughs's family friend, and supposedly had given the manuscript to Burroughs with instructions not to publish it for 21 years.

The story went something like this. John Carter was a captain who fought for the Confederacy in the American Civil War. He was a strapping man, 6'2" and every bit the iconic hero. He reveals in the book that he has no memory of childhood, always having been 30 years old. People have grown old around him, yet he never ages.

Carter was mustered out of the Confederate army and joined forces with a military buddy and began prospecting for gold in the part of the country that later became Arizona. After striking it rich, he and his companion were attacked by Apache Indians, and his friend was killed. Carter retreated into a cave, where he was overcome by fumes and apparently died. And then the fun began.

Carter awoke in Barsoom, which is the native term for the Red Planet. Burroughs's Barsoom will be familiar to those who recall Percival Lowell's Mars. A million years ago, Barsoom was a lush place, covered with oceans. However, in the intervening years, the water evaporated, lost to space. Barsoom was a dying planet, dry and sandy. The residents worked feverishly to build canals to bring water from the polar ice caps to the equatorial regions, trying desperately to keep alive.

The residents of Barsoom were not only humanoid but very much like *Homo sapiens*, except for being oviparous. They had navels—in spite of laying eggs—and breasts, assuring the eye-catching covers that were typical of the pulp serials at the time. They lived at least 1,000 years, although it is possible that they lived longer. Their culture required that when Martians reached that age, they travel down the River Iss. This trip nominally brought them to paradise, although, as we will see, the trip was considerably less pleasant than that.

Martians came in different colors and with different temperaments. They were red, green, yellow, white, and black, and their politics were generally either theocratic or dynastic. Red Martians dominated Barsoom, although this doesn't mean they had a single, global government. Instead, they were organized into several competing city-states, with the polity Helium being of particular import in the first book. Red Martians were highly civilized, with a strict code of honor. They respected personal property, formed families and strong alliances. Their technology was advanced (especially compared with Burroughs's Earth) and included flying machines, both civilian and heavily armed warcraft. Barsoomian scientists had mastered genetic engineering, medical transplant techniques, faxes, and television and had incorporated radium into their long-distance weaponry. Radium was discovered in 1898 and first extracted in metallic form in 1910, so this was cutting-edge stuff.

Red Martians were bred as a hardy race that would help them survive the increasingly dry conditions of Barsoom. They were a mixture of the Yellow, White, and Black Martians, all of whom had nearly died out. While the Martians had mastered advanced technology, they preferred to fight hand to hand with swords and comparable weapons. This makes the descriptions of the battles exciting and vivid.

Green Martians can be visualized as barbarians. The males are 15 feet tall and the females are 12 feet tall and appear to be the outcome of a genetic experiment gone awry. They were nomadic and warlike. Any enemy captured by a Green Martian was tortured, frequently to death. Rising in the social structure always involved battle, and becoming a leader of the various warring tribes could only be attained by winning mortal combat. There was no family structure in Green Martian civilization; allegiance was only to the tribe.

Yellow Martians lived in a few small and domed cities near the North Pole. They were exceptionally cruel and used a tractor beam to pull down aircraft so they could enslave the crew. They show up rarely in the series.

White Martians were once the master race of Barsoom. They were once thought to be extinct, but various isolated populations were revealed over the course of the eleven books. One population called the Lotharians had evolved to be reclusive intellectuals who lived separately from all other Martians and spent their time debating philosophy. Another group called the Therns inhabited the Valley Dor, which was the terminus of the River Iss. The valley was actually populated by vicious creatures controlled by the Thern. These creatures typically killed Barsoomians taking their journey to paradise and ate their flesh.

Black Martians inhabited a hidden fortress near the South Pole. They called themselves "the First Born" and considered themselves to be unique among the Martians. They sometimes raided the Thern, but they don't show up often in the series.

The plot of Barsoom tales are hardly complex. A male hero, noble and brave, is forced to travel to a faraway place to rescue a woman he loves. The woman has been captured by a powerful man who desires her both sexually and as a means to make political gains. Along the way, the hero encounters many adventures: battles, captures, escapes; action is the norm and subtlety rare.

Burroughs's Barsoom stories had many themes that would have resonated with his early twentieth-century readers. A civilized, European/American hero enters a barbaric world, which could easily be related to Kipling's stories of West Asia or tales of the conquest of Africa or the American West. By 1912, the era of frontiers was fading, and Americans were beginning to romanticize their history. The term "spaghetti western" was more than half a century in the future, but Burroughs's readers would have appreciated Clint Eastwood movies, with a flawed, but fundamentally brave and noble protagonist, a distinctly evil villain, and a brave, but vulnerable, heroine. A lawless world, inhabited by hard men, and a hero who must live by the local rules to survive; the Man with No Name or John Carter of Mars, the tale is a familiar one.

Another dominant theme in the Barsoom tales is the one of race, which was certainly one to which the readers would be receptive. Just fifty years after the American Civil War, the bulk of readers would have had no difficulty accepting the concept of superior and barbaric races. The era of European colonialism was waning and underwent dramatic change in the aftermath of World War I, which would begin just two years after *Under the Moons of Mars* was published. The Barsoom series was published through 1943, so it is natural that a series in which differently colored Martians were featured so prominently, each with their own characteristic racial identity, would resonate with Americans who were wrestling with their own racial difficulties.

Burroughs's Barsoom series had an indirect impact on the public's view of Aliens. It never received the publicity of Wells's work, but it was a tremendous influence on subsequent science fiction writers. Ray Bradbury, Isaac Asimov, and many other famous science fiction authors grew up reading of Barsoom. James Cameron has stated in interviews that his highly successful *Avatar* movie was inspired by Burroughs. The popularity of the Tarzan comic strips led to a short spate of Barsoom-oriented comics in the Sunday papers in the early 1940s. And, of course, the 2012 release of *John Carter*, a big budget picture bankrolled by the Disney Company, has introduced a whole new generation of viewers to Barsoom. The commercial success of this movie was disappointing to the film's producers, but it is possible that Burroughs's impact on the public might increase.

The Pulps

Science fiction has gone through many phases over the years. From the 1920s through the late 1940s, the most common form of science fiction was in magazines. As we have seen, Edgar Rice Burroughs's first novel was in serialized form. However, the magazine in which he published it was not a science fiction magazine. The first magazine devoted to science fiction, *Amazing Stories*, had its initial publication in April 1926 and was edited by Hugo Gernsback. Gernsback's contribution to science fiction has been acknowledged through lending his name to the prestigious Hugo Awards, established in 1953.

Amazing Stories was published, with some interruptions, for about 80 years. Shortly after the magazine was started, the readership soared to 100,000, although by 1938, it was down to only 15,000. Over the decades, the periodical went through many editors, many publishers, and many visions. Even though it is now acknowledged as the first science fiction magazine (indeed before the term science fiction was even coined), *Amazing Stories* was soon eclipsed as the leading periodical in the genre.

If *Amazing Stories* was the vanguard of the science fiction revolution, the flagship was *Astounding Stories of Super Science*, which began operations in 1929 and continues today. The name of the magazine underwent many changes over the years and is now *Analog: Science Fiction and Fact*. Fans refer to the magazine as simply *Analog*. The onset of John Campbell's tenure as editor in late 1937 is considered to be the start of the golden age of science fiction, a period that ran until the mid-1950s, at which time Campbell's strong personality alienated some of his best writers, and they started publishing in other magazines. Plus, as we will see, the science fiction environment changed in the early 1950s.

Still, *Analog* introduced to its readers fledgling authors who became some of the most famous writers of science fiction, with L. Ron Hubbard (later the founder of Scientology), Clifford Simak, L. Sprague de Camp, and Henry Kuttner (one of my favorites) and his wife, C. L. Moore. Other new authors who grew to greatness in its pages were Lester del Rey, Theodore Sturgeon, Isaac Asimov, A. E. van Vogt, and Robert A. Heinlein.

The pulps were not held in high regard by the parents of the magazines' many adolescent readers. Known for lurid covers that frequently featured brass-bikini-clad women in imminent danger of being eaten by a monster or an Alien of one kind or another, the pulps led many a parent to make a disparaging remark about the quality of the literature. While some pulps did try other cover art, including some that was a bit more serious, those issues inevitably sold much more poorly than the ones that showed a lot of skin. Then, as now, sex sells, and sex and danger sell even better.

Amazing Stories and *Analog* were by no means the only science fiction magazines out there. Over the years (and especially in the 1930s and 1940s), more than a hundred different magazines were published in this genre and that doesn't include their cousin pulps, the horror magazines.

However, while the pulp magazines were very popular among the diehard science fiction audience, they were not considered serious literature and had a relatively small direct impact on the public. Serious people did serious things and certainly didn't spend their time reading spectacular hoo-hah, although many budding scientists certainly enjoyed the pulps when they were young.

Flash and Buck

In order to make a greater impact on the public, science fiction writers had to exploit other media. The big ones in the period of 1920 to 1940 were newspapers, radio, and newsreels. Among the first forays of science fiction into these venues were Buck Rogers and subsequently Flash Gordon.

Buck Rogers was introduced in *Armageddon 2419 A.D.*, published in the August 1928 issue of *Amazing Stories*. The story was translated into a syndicated comic strip in January 1929 as *Buck Rogers in the 25th Century A.D.*, by pure coincidence, the same month that *Tarzan* began as a newspaper comic. While the original Buck Rogers article told the tale of warfare on postapocalyptic Earth, the stories expanded over time. By the 1930s, short films were made, including one for the 1933–1934 World's Fair called *Buck Rogers in the 25th Century: An Interplanetary Battle with the Tiger Men of Mars*. Serialization was soon to follow.

If Buck Rogers was the first in this genre, Flash Gordon led the way into the world of Aliens. Flash Gordon was introduced to the public as a hero in a science fiction comic strip that began in January 1934. The comic was inspired by the earlier and successful Buck Rogers strip and was intended to compete directly with it. When the Earth was bombarded by meteors, Flash

Gordon and his companions Dale Arden and Doctor Hans Zarkov set out to investigate. Zarkov invented a rocket that allowed them to head into space to determine the meteor's origin. Originally Zarkov kidnapped Flash and Dale, but Flash quickly became the leader.

The meteors originated from the planet Mongo, which was led by the despotic and cruel Ming the Merciless. Ming, although an alien, was essentially human with flamboyant dress and classical Persian (Iranian) features of dark skin and a dark and neatly trimmed beard. Fans of the original *Star Trek* series would recognize Ming as looking like the classic (i.e., original series) Klingon. While Ming the Merciless is the most famous enemy of Flash Gordon, the three companions travelled Mongo for years, encountering the Shark Men, the Hawk Men, and the Lion Men.

Flash Gordon was also serialized on radio in April 1935 in *The Amazing Interplanetary Adventures of Flash Gordon*, which was an adaptation of the comic strip. Three film serials were created starring Buster Crabbe: *Flash Gordon* (1936), *Flash Gordon's Trip to Mars* (1938), and *Flash Gordon Conquers the Universe* (1940).

In the 1930s and even later, movie serials were short films, perhaps 10 minutes long, that told a piece of the story and ended with a cliffhanger. The following week would show the next installment of the story. Moviegoers would attend a movie and see a couple short films, including newsreels, followed by the week's main attraction or maybe a double feature. In a world in which there was no television, people would go to the movies for entertainment. Even if they weren't interested in Flash Gordon or Buck Rogers, they would see the serial. Through these newspaper comics, film serials, and radio shows, science fiction was being introduced to the general public.

The 1930s were a dark time for the world. The stock market crash of 1929 had signaled the beginning of the Great Depression. This was followed by a decade of war. Times were harsh, and the movies of the time were used for escape. Science fiction was pure escape, adventures without any real connection to the real world.

Aliens and the Iron Curtain

During World War II, people were focused more on defeating the Germans and Japanese than they were questions of outer space. Some people's priorities remained unchanged, as then-radar instructor and eventually leading science fiction author Arthur C. Clarke bemoaned the failure to ship his beloved issue of *Analog* (then *Astounding Stories*) from the United States to Britain, "owing to the war, regular supplies of *Astounding Stories* had been cut off by the British authorities, who foolishly imagined that there were better uses for shipping space."

However, during the war, mankind had heard about foo fighters. Given the hundreds of stories in the press about flying saucers in the late 1940s, it was entirely natural that imaginative Aliens would begin to appear in the public eye. Science fiction was about to become far closer to the mainstream.

The late 1940s and 1950s were a time of economic strength but also of considerable uncertainty. In March of 1946, Winston Churchill gave his "Iron Curtain" speech, "From Stettin in the Baltic to Trieste in the Adriatic an 'iron curtain' has descended across the continent . . . in what I must call the Soviet sphere, and all are subject, in one form or another, not only to Soviet influence but to a very high and in some cases increasing measure of control from Moscow." The era of Red Fear had begun. Let's spend a little while thinking about the world in which people of the 1950s lived.

The German and Japanese empires had been defeated. But mankind now lived in a nuclear age, in which a single bomb could incinerate a city. In 1952, the United States detonated the first fusion bomb, with a yield about five hundred times more powerful than the Hiroshima and Nagasaki bombs. The Soviet Union detonated its first fission bomb in 1949, followed by a fusion bomb in 1953. The two major power blocks on the planet had unleashed the power stored in the nucleus of the atom and could kill a million people in an instant. Were these two great powers allies or enemies?

Well, it's not quite fair to call the Soviets and the Americans enemies, but they certainly were rivals and potential combatants. Diametrically opposed political and economic viewpoints (and more than a little self-interest) guided their thinking, and the propaganda of both sides painted the other as an evil enemy, just waiting to invade and destroy the people that had elected to follow the "right" way of life. The year 1948 saw the Berlin Blockade, while 1950 brought the proxy war in Korea. Also in 1950, an undistinguished Wisconsin senator named Joseph McCarthy made the blockbuster statement in a speech, "While I cannot take the time to name all the men in the State Department who have been named as members of the Communist Party and members of a spy ring, I have here in my hand a list of 205." For the next several years, American politics was dominated by a witch hunt. People were accused of being communist sympathizers, lives were ruined, and the Red Menace was seen to be everywhere.

Classic Alien Films of the Fifties

So, outside and hostile infiltrators could be anywhere and everywhere. An atomic war could vaporize millions and a flying saucer craze was a recent memory. These problems were in the back of the mind of the audience who experienced the world of 1950s science fiction.

And what a world it was. The 1950s brought dozens and dozens of Alien movies. Many "B-quality" flying saucer and invasion movies appeared at the time and have long since been forgotten. A few were iconic and are still remembered today. We will talk about some of these in order of the year that they were released.

The Day the Earth Stood Still

One of the first of the Alien movies of the 1950s is *The Day the Earth Stood Still* (1951), which was a cautionary tale about the dangers of nuclear war. The movie starts out quickly, with a radar blip circling the Earth at high altitudes at a speed of 4,000 mph. The opening sequence has word of the high altitude object being spread across the world essentially immediately, with scenes being shown of radio announcers in India, Great Britain, the United States, and others telling of the observation. The U.S. announcer says, "This is not another flying saucer scare. Scientists and military men are in agreement that, whatever it is, it's real." We should keep in mind that radar was less than a decade old, having been used in a military environment in World War II and, in addition, the flying saucer frenzy of 1947 had occurred just a few years before. This was a timely and high-tech touch to the movie.

The radar blip closes on Washington, D.C., and is revealed to be a classic flying saucer, flat on the bottom, with a smooth upper curve, like a squashed bell. The saucer is the color of brushed aluminum, and it lands on a park field dotted with baseball diamonds near the Washington Mall.

In an unrealistic display of governmental organization and efficiency, the saucer is surrounded unbelievably quickly with tanks, antiaircraft guns, and troops, setting up the first drama. Two hours after landing, the saucer opens and a humanoid figure walks out, dressed in a jumpsuit. An antsy soldier pulls a trigger and a shot rings out, hitting the Alien in a shoulder.

As the Alien lies on the ground, another figure appears in the door of the saucer, an ominous eight foot tall silver robot named Gort. Gort has a visor, which can open and from which a laser-like weapon can shoot. Gort shoots at rifles, an antiaircraft gun, and a tank with his beam, disintegrating them all.

By now, the Alien, who introduced himself as Klaatu, has been helped to his feet. He stops Gort from doing further damage, and the robot seems to turn off, going into some sort of sentry mode. As Klaatu is brought to a hospital, the door of the saucer closes, sealing the inside from prying human eyes.

In the hospital, the Alien talks with a representative of the U.S. president, telling the representative that he needs to speak with all the leaders of the world, but he is told that it is highly unlikely that this can be arranged. When Klaatu insists that his message is too important to be given to just one group, the representative tells him that "our world is full of tension and suspicion." The real-world Cold War is reflected in the film.

Klaatu then escapes from the hospital, stealing a suit that allows him to blend into the population. He rents a room in a boarding house, befriending a young widow of World War II and her son. Over the next days, Klaatu determines that the smartest man alive is a physics professor and manages to arrange a meeting with him. During the meeting, Klaatu identifies himself and again asks for help in arranging a meeting with world leaders, but the professor isn't confident of his ability to do so, noting that scientists are often ignored. Klaatu tells him that if he doesn't speak with the leaders, "the Earth is in danger of being eliminated." They agree that Klaatu will arrange some sort of nonlethal demonstration of his power. He does so by disabling electricity across the entire planet for a half hour, except for things like hospitals and planes in the air.

The meeting with scientists is arranged but, on the way to the meeting, Klaatu is killed. Before he dies, he gives the woman he has befriended a phrase that she must tell the robot Gort. When she arrives at the saucer, Gort has awakened and has already killed two soldiers guarding him. As Gort advances on the woman, she utters one of the most famous phrases of movie science fiction, "Klaatu barada niktu." This phrase is never translated in the movie but seems to be some sort of "safe phrase." Gort reacts by taking the woman into the saucer. Gort then retrieves Klaatu's body and temporarily revives him from the dead. Klaatu tells the woman that the revival is only temporary, as the power of life is reserved for a "higher spirit."

The movie comes to its dramatic conclusion, with Klaatu exiting the saucer with the woman and Gort as guard and addressing the assembled crowd (figure 3.2). He tells them that the Earth's internal affairs are our business, but, if we take our wars into space, that the community of alien worlds will then take action. The alien community has built the robots as policemen of the cos-

FIGURE 3.2. In the closing scene of the movie *The Day the Earth Stood Still*, Klaatu and the robot Gort stand on their saucer and warn mankind of the dangers of bringing the Earth's conflict into space. 20th Century Fox.

mos and that power cannot be taken back. The band of civilized space-faring races has given up weapons and war, knowing that the reaction of the robots would be immediate and terrible. Klaatu closes the movie with, "I came here to give you these facts. It is no concern of ours how you run your own planet. But if you threaten to extend your violence, this Earth of yours will be reduced to a burned out cinder. Your choice is simple. Join us, and live in peace or pursue your present course and face obliteration. We shall be waiting for your answer. The decision rests with you."

The story here is simple and clear. Mankind of 1950 was on the precipice of imminent annihilation. The Soviet Union and the United States both had fission weapons and were vying for world domination. Incineration of civilization is a real and pressing concern. The world had just completed the most horrific war ever, in which many tens of millions of people had died. The gray specter of communism was a real danger and the memories of world war only too fresh. It is unsurprising that the movie reflected these worries. It is also interesting to see that the worry of communist infiltration was not a central theme, except for when one matron alluded to the idea that the flying saucer was a Soviet creation. The full impact of McCarthyism was still a future worry. This movie was remade in 2008.

The Thing from Another World

The 1951 movie *The Thing from Another World* took some of the iconic filmmaking character types from earlier monster films and brought them to science fiction. These now recognizable types include the hard-headed and suspicious military man, the naïve and arrogant scientist, the reporter concerned only with getting a story, and an isolated group, with no hope for reinforcements. The movie is much faster paced than most others of the era, with an intensity that would not be out of place in a modern movie. Part *Frankenstein* and part *Alien*, it was patterned on John W. Campbell's 1938 novella *Who Goes There?* One major difference between the film and the original is that in the novella the Alien is a shape shifter.

The story begins with an air force flight crew investigating a report of a plane crash near the North Pole. With a reporter accompanying them in case there is an interesting story, they fly to the remote Polar Expedition Six, the research site of a brilliant Nobel Prize–winning scientist. It is determined that the object that crashed was metallic, too large to be a plane and too nimble to be a meteor.

When the air force crew and some of the scientists arrive at the impact site, they find a large spot where the polar ice had been melted and frozen over (figure 3.3). A single rudder made of some unknown alloy sticks out of the ice. The airmen and scientists spread out at the edge of the object and realize that they are standing in the shape of a circle. One of the airmen exclaims, "Holy cow! We found one! We found a flying saucer!" Recall that this movie was released just four years after the flying saucer frenzy.

The captain decided to melt the ice using thermite. In a moment of Hollywood hyperbole, they claimed that a single thermite bomb would melt hundreds of tons of ice in 30 seconds. To their horror, the thermite ignites the skin of the flying saucer, and it is utterly destroyed. (This isn't explained in the movie, but in the original novella, it was explained by having the flying saucer made of magnesium.)

The destruction of the saucer seems like a disaster, but clicks on a Geiger counter lead them to an 8 foot tall body frozen in the ice. With a storm on the way, the airmen chop out a cube of ice with the body encased in it, load it on the plane, and fly back to the base. During the flight, the filmmakers pay homage to the recent UFO frenzy and the air force's famous response, known as Bulletin 629-49. An airman reads "from the Office of Public Information, December 27, 1949, Bulletin 629-49, regarding item 6700, extract 74,131: The

FIGURE 3.3. The Thing's spacecraft has melted the ice and been frozen underneath in this image from *The Thing from Another World*. In the left photo, we see the rudder of the craft, while in the right picture, we see the airmen standing around the perimeter of the craft, showing its circular shape. Winchester Pictures Corporation.

air force has discontinued investigating and evaluating reported flying saucers on the basis that there is no evidence for their existence. The air force said that all evidence indicates that reports of unidentified flying objects are the results of: one, misinterpretation of various conventional objects; second, a mild form of mass hysteria; third, that they're jokes." While the movie text is not a direct quote of the real bulletin, it is similar in message.

When they make it back to the polar base, tension sets in between the lead scientist and the captain. The scientist insists on melting the ice to get at the creature, while the captain, wary after the destruction of the saucer, instead insists on keeping the Alien frozen until he receives orders. Given that the captain has troops and guns, the military mindset wins. The block of ice is brought into a storeroom, and the windows are broken to ensure a frigid environment. Sentries guarding the creature are, of course, very cold and are given an electric blanket to make their time more comfortable. They are unnerved by the creature's eyes, and one of the guards covers them with the blanket without thinking through the combination of a heated blanket and ice. The dramatic tension mounts as the audience sees the water drip off the melting block.

Apparently freezing doesn't affect some Aliens, as, once the block melted enough, the creature revives and comes toward the guard. The guard panics and shoots it several times with a pistol and runs away. The bullets seem to do no damage and, when the guard returns with more men, the creature is gone. However, they hear noise outside and see something fighting with the sled dogs. In a flurry of blows, the Alien flings the broken bodies of the dogs in all directions and then escapes. When the men check the carnage, two dogs are dead and one is missing, but they find the creature's arm and heavily clawed hand.

The next scene opens with the scientific team poking at the hand. They determine that the creature is not an animal. It is a vegetable, and they even find seedpods in its arm. The reporter has difficulty believing that the creature was a vegetable and comments that "it sounds like you're talking about a super carrot." The station's lead scientist retorts with "this carrot, as you call it, constructed an aircraft capable of flying some millions of miles, propelled by a force yet unknown to us."

Meanwhile, the chief scientist has surreptitiously taken the seedpods and "watered" them with human plasma. A plant begins to grow. There is a disagreement between the scientists on whether this course of action is wise, however the chief scientist tells them not to let the military know about it. He is adamant that they should contact the creature, not kill it. In a dramatic moment, an incoming radio transmission from an air force general forbids destroying the Alien.

The captain believes the orders to be foolish, so he discusses with his men how to kill the creature. As the men discuss among themselves how to kill a vegetable, a female character suggests boiling, baking, stewing, or frying. (After all, this was the 1950s, and men weren't expected to cook.) This advice leads them to decide that fire was the way to go. They rummage up some kerosene however, as they do it, the Geiger counter again starts to click. The creature is returning, coming (literally) for blood. It breaks through a door and is seen in silhouette, looking essentially like the classic Frankenstein's monster. The airmen throw kerosene on it and set it afire. The creature escapes by jumping through a window and dousing the flames.

Since fire seemed to be only a marginal deterrent, the airmen decide that electrocution might be more effective. While they start to put their plan into action, the creature shows that it is intelligent by cutting off the fuel to the heaters. Knowing that the next target would be the generators, they decide to make a last stand there and set up the electrical trap.

When the Alien attacks again, it is destroyed by lightning and reduced to a smoking mass. The airmen then burn all the scientist's notes and the seedlings that are still in the lab. Nothing is left to prove the creature even existed.

The bad weather begins to lift, allowing radio communications to resume. After spending the whole movie moaning about not being able to submit his story, the reporter is finally given the microphone to tell his tale. The movie ends with his report, in which he says, "And, before giving you the details of the battle, I bring you a warning. Every one of you listening to my voice, tell the world, tell this to everybody, wherever they are. Watch the skies. Everywhere keep looking. Keep watching the skies." After all, where there is one flying saucer, there could be many.

The movie *The Thing from Another World* taps into many phenomena. The flying saucer frenzy of 1947 was a recent memory, as were the atomic explosions ending World War II, with the subsequent development of a Soviet weapon. Atomic energy was thought to be the future, but the power and danger of atomic weaponry was well known. It isn't surprising that filmmakers would invoke radioactivity as something that would identify the Alien. In addition, the Alien arrived in a flying saucer. Like the much later *Alien* movies, the creature from outer space was powerful, invincible, and totally alien. Mankind had little in the way of defenses in a one-on-one battle, making human beings the prey. It's a classic monster story with an extraterrestrial twist. The movie was remade twice, first in 1982 and again in 2011.

Red Planet Mars

Red Planet Mars (1952) is the Alien film that perhaps most overtly involves communism. Even the title is a clearly intended double entendre between the planet and the Soviet Red Menace. A Nazi scientist invents a "hydrogen valve" that makes it possible to boost a radio's performance sufficiently so as to contact Mars. He is sent to prison by the Soviets before he can make the device. An American scientist finds the plans in the Nuremburg files and makes such a radio, determined to communicate with Martians. He is certain that there is intelligent life on Mars because an astronomer observes canals on Mars, along with a huge ice cap. A second observation of Mars shortly thereafter reveals the ice cap has melted and the canals are full of water. We see Percival Lowell's influence a half century later (figure 3.4).

The American scientist has difficulty communicating with the Aliens until his son suggests using the value of *pi*. (His son comes up with this idea while eating the last piece of pie.) The scientist sends out 3.1415. When the Martians send back 3.1415926, the scientist is certain that communication is possible. Martians tells the scientist that they live to an age of 300 years, can feed a thousand people on a half acre of land, and get power from cosmic rays. This news upends the economies of Western civilization.

Meanwhile, we find the Nazi scientist in the Andes, having replicated his

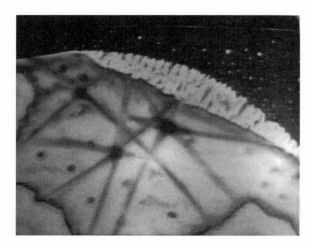

FIGURE 3.4. *Red Planet Mars* (1951) clearly shows the ongoing impact of Lowell's Martian canals (see figure 1.2) , even decades after they were scientifically discredited. In this image from the movie, we see the canals and the polar icecap that was subsequently melted and the canals filled. Melaby Pictures Corporation.

earlier equipment using money and supplies from the KGB. He is unable to contact Mars himself but can monitor the American's transmissions. The KGB finds him and threatens him if he is unsuccessful in his own attempts to contact Mars.

Things get weirder when Martians speak of their philosophy, going so far as quoting the Christian Sermon on the Mount. This has a huge impact on Western society, but an even bigger one behind the Iron Curtain. A resurgence of religion in the Soviet Union brings down the government.

Later, the Nazi scientist makes his way to the American's laboratory to reveal that it wasn't Martians with whom the Americans were communicating but rather the Nazi using his equipment in the Andes. His motivation was to bring down the governments of the West and of the Soviet Union. However, it turns out that he only made the economic communications. The philosophical and religious ones actually *had* come from the Martians. In a bit of a muddled plot twist, the laboratory is exploded by a gas leak from the hydrogen valve that is shot by the German scientist.

Red Planet Mars does not show Martians and is uncharacteristic in that it portrays them as peaceful people. Many other films of the era depicted Martians as invaders or at least potential adversaries. With the exception of communication with Mars as a plot device (and the nod to Lowell's impact), the movie is more about the earthly cares of the 1950s than about Aliens.

Invaders from Mars

Invaders from Mars (1953) is much more subtle in its depiction of the era's worries and introduces the ideas of infiltrators: people who look familiar but are clearly enemies. A young boy sees a glowing flying saucer land in his backyard in the middle of the night. His father goes out to see what he's talking about and doesn't come back until the morning. When the father returns, he is a changed person. We don't know it yet, but the Martians have implanted an electrode into his neck that controls his behaviors.

Several additional people are eventually altered. We see this indirectly, as the saucer is now buried under a sand pit and when people walk over the sand, a hole appears and they fall underground and are presumably captured. The boy is unable to convince anyone that something is going on until an attractive female psychologist and her male astrophysicist friend listen to him. It turns out that the astrophysicist is aware of recent flying saucers being spotted on radar. They speculate that the Martians might be invading as a preemptive strike against Earthlings starting to develop rockets, and so they contact the army.

Eventually the army is able to establish that there is a saucer under the ground and they break into it with a commando team. Naturally, the little boy and the psychologist also end up in the saucer and are kidnapped by Aliens, which are large, lumbering humanoids. The humanoids are perhaps not sentient, as they seem to respond to the direction of a different form of Alien, one that can be described as a rather human-like head, set on top of a mass of tentacles. This Alien lives in a big, clear globe and is carried by the lumbering Aliens (figure 3.5). This Alien is interesting in that it isn't strictly humanoid, although it clearly has human-like features. Ironically, these human-like features drive home the point that it is an Alien leader and a huge head indicates its intelligence. Had it looked like a tree, a squid, or a lump of moss, the audience would have had a harder time identifying it as an Alien.

The commando team finds the psychologist and boy and everyone escapes but not before setting demolition charges. The Alien attempts to escape by flying away, only to have its saucer explode in midair.

Invaders from Mars shows the paranoia of the time of not knowing who among your neighbors might be an enemy. It also showed humanoid and human-derivative Aliens and a classic flying saucer of the sort that had ap-

FIGURE 3.5. The dominant Alien from *Invaders from Mars* has a large head, indicative of intelligence and possibly telepathic control over the lumbering worker Aliens. He has a small body and tentacles. Note the strings to make the "hand tentacles" wiggle. National Pictures Corporation.

peared in the press just a couple of years before. The saucer reports in the newspapers had told the screenplay writers what an Alien ship should look like. This movie was remade in 1986.

Forbidden Planet

Forbidden Planet (1956) has been called "Shakespeare's *Tempest* set in space," but it could easily have been a pilot show for the 1966 television series *Star Trek*. A spaceship had landed some years before on Altair 4 and communication was lost. Earth sent a military spacecraft, the United Planets Cruiser C57-D, to check it out. This spaceship is a classic UFO, exactly as you'd imagine a flying saucer to look, except it is a ship from Earth (figure 3.6).

When they arrive at Altair 4, the military team makes radio contact with the surface, where they are warned away. The planet is dangerous. Ignoring this advice, the saucer lands (no transporters) and the captain, first officer, and doctor are driven by a robot named Robby, who takes them to meet the only surviving passenger of the first spaceship, a brilliant scientist, along with his beautiful daughter, who was born after leaving Earth. He tells them how something killed all the other passengers and blew up the original spacecraft when it was trying to escape.

After some romantic tension between the beautiful daughter and the cap-

FIGURE 3.6. The United Planets Cruiser C57-D from the movie *Forbidden Planet* is a textbook example of a flying saucer. When it lands it sits on three legs that double as staircases. Metro-Goldwyn-Mayer Studios.

tain, it is revealed that Altair 4 was once inhabited by a race called the Krell. The Krell's technology was far superior to anything developed by mankind, although they had been somehow destroyed by their technology in a single night, more than two thousand centuries before. However the scientist had figured out how to work some of their equipment that he had found.

Meanwhile, something is killing the crew. Eventually, a large flame-like creature tries to get through the C57-D's force fields. The captain decides to evacuate the planet and returns to collect the scientist and his daughter. The scientist's compound is attacked, and it is eventually revealed that the monster is the unconscious mind of the scientist, amplified and personified by the immense power of the Krell. The scientist is overcome by remorse for what he did to the passengers on the original spaceship, so he sets a self-destruct sequence that the Krell had put into place and then dies. The captain is able to escape with the daughter and Robby and the C57-D makes it just far enough away to avoid being destroyed by the detonation that shatters the planet.

Fans of *Star Trek* will recognize *Forbidden Planet* as a story that follows a typical *Star Trek* plotline. Gene Roddenberry, creator of *Star Trek,* admitted in his official biography that *Forbidden Planet* was one of the inspirations for his popular television series. For our purposes, we never encounter the Krell, except that we know that they were an ultra-powerful race, undone by their own technology and hubris. And the C57-D is an iconic example of a flying saucer, cementing in the audience's mind the image that originated in a headline writer's creative interpretation of Kenneth Arnold's first UFO report.

Earth vs. the Flying Saucers

Earth vs. the Flying Saucers (1956) is another example of the presence of flying saucers in the public mind. As the title suggests, this movie was a classic shoot 'em up between Earth and Aliens. The movie starts with a newlywed couple driving on a deserted highway, when they were buzzed by a flying saucer. This is somewhat reminiscent of Betty and Barney Hill, although the movie preceded their experiences by five years, and no amnesia was involved. The husband is the lead scientist of a government program called Project Skyhook, which involves launching artificial satellites into space, and he is driving while dictating his notes into a tape recorder. As the saucer buzzes the car, it emits a high-pitched insect-like sound, which is recorded on the tape.

The two continue to the base from which the satellites had been launched. After being informed that the previous ten satellites had been shot down, the scientists launch the eleventh satellite, which also gets lost. Shortly after the launch of this last satellite, a flying saucer lands at the base, and three human-oid Aliens come out. One of them is shot by a soldier, and the Aliens retaliate with ray guns that kill. The flying saucer then devastates the base hosting Project Skyhook. The husband and wife become trapped underground, and, as the air grows stale, he records what he thinks will be his last words. As the batteries die, he plays the sound the saucer makes, only to find out that it was a sped-up verbal notification that the Aliens would be landing at Project Skyhook.

The husband and wife are rescued and go to Washington, D.C., where they tell their story. The Aliens radio the scientist how to contact them and, against orders from U.S. officials, he goes off to meet them. His wife notifies an army major who had been detailed to watch them, and the two follow the scientist driving at high speed. The scientist is joined by his wife and the major, and the three enter the saucer. The Aliens reveal that they intend to take over the world and show that they can read peoples' minds and learn everything the person knows.

The scientist is given 56 days to bring together the leaders of the world to Washington, D.C., to surrender to the Aliens. Then the scientist, his wife, and the major are released. The scientist builds two weapons, first a sonic ray of minimal destructive power and then an electric beam that disrupts the flying saucer's magnetic propulsion system. Many versions of this electric beam are built and installed in army trucks.

On the appointed day, the flying saucers approach the U.S. capital and

FIGURE 3.7. In the movie *Earth vs. the Flying Saucers*, a fleet of classic flying saucers wreak havoc on Washington, D.C. Here they fly over the Lincoln Memorial. Clover Productions.

start blowing stuff up (figure 3.7). The earthlings fight back and eventually shoot down the saucers. It seems that the battle scenes of the much later movie *Independence Day* might owe a creative debt to *Earth vs. the Flying Saucers* in how the Aliens damage Washington, D.C.

Earth vs. the Flying Saucers is not an overtly political movie, but it is a classic conflict between good guys and bad guys. The flying saucers are quite stereotypical in design and demonstrate yet again how deeply the saucer image had permeated the culture. Aliens and flying saucers were now solidly mainstream.

Wrap Up

The glut of 1950s UFO movies began to taper off before the end of the decade, shortly after the Soviet launch of Sputnik in 1957. The desire for imaginative depictions of flying saucers was over, replaced with the very real earthly space race. Mankind realized that we could conquer space ourselves, and the idea of talking about Aliens in space somehow seemed passé. In the next chapter, we will discuss the period from 1960 to the present. Even though this spans more than half a century, the character of science fiction had changed. The era of the high-impact blockbuster had come. From the period of the 1970s onward, science fiction was a mainstream staple in the television and movie industry. Aliens were everywhere, although not always intended to be taken seriously. The filmmaking and moviegoing culture had changed to consciously treating Aliens as a way to express our own, earthly, concerns. Essentially, Aliens were no longer alien. Instead, the past few decades have resulted in a handful of high impact movie and television franchises that have extensively infiltrated the psyche of humanity. In the next chapter, we will talk about them.

FOUR

BLOCKBUSTERS

But please remember: this is only a work of fiction. The truth, as always, will be far stranger.

Arthur C. Clarke, *2001: A Space Odyssey*

In the 1950s dozens of movies were made that reflected the influence of the flying saucer craze that began in 1947 and the worries that came with the beginning of the Cold War. Blockbuster movies of recent decades have a different flavor. Accompanied by huge advertising budgets and professional marketing campaigns, this new type of science fiction film began to shape the public's view of Aliens. In recent years, the ubiquity of cable television and the need to fill hundreds of channels has led small budget science fiction film producers to easily connect with the niche audience of science fiction enthusiasts; for instance the ungrammatical SyFy network shows newly made movies that would have made Ed Wood blush. (If you're not a fan of the genre, Ed Wood was notorious for making very bad science fiction movies.) However, these new "B-movies" have but a modest impact on the public. Most of us saw modern-day Aliens on the big screen. In this chapter, we'll discuss the high impact movies and television shows of the past few decades.

The 1960s were a relative desert for movies about Aliens of any kind. The biggest movie of that decade was the artistic *2001: A Space Odyssey* in 1968, which only hinted at Aliens, in that they created obelisks that keep popping up in the movie. The first one appears in front of a group of plant-eating and timid hominids and alters them. They kill a rival group's leader and this is the start of the lineage ending with *Homo sapiens*. The movie then jumps ahead to 2001 when an obelisk is found on the moon. When explorers get to the lunar obelisk, it sends a signal to Jupiter. The rest of the movie details a journey to Jupiter with a recalcitrant computer named HAL. Another obelisk is found near Jupiter, and, when activated, it brings an astronaut through a time sequence of aging and a tremendous light show, ending up with his being enclosed fetus-like in an orb of light staring at Earth. The idea is conveyed that perhaps mankind has again been altered and is entering a new stage of development.

The Aliens in *2001: A Space Odyssey* are never shown. This is in part due to advice from Carl Sagan, who suggested to the film's creators that a realistic Alien would not be humanoid. Given the constraints of movie making of the era, in which computer graphics were quite primitive and therefore Aliens would need to be shown using human actors, the directors opted to not show the Aliens at all. This technique was copied in the 1997 movie *Contact*, penned by Carl Sagan and his wife, Ann Druyan. Sagan was famous for reminding people that Aliens won't look like us. *Contact* had religious overtones but also told a tale of Aliens that were so far advanced from us as to appear godlike.

The rest of the 1960s really was a low point in film depictions of Aliens. This was the era of Betty and Barney Hill, and it would take some time for these new ideas of Aliens to trickle into Hollywood. Television was another matter. The 1960s saw such series as Britain's long-running *Doctor Who* and cult classics *Lost in Space* and *The Twilight Zone* (which occasionally featured Aliens). But arguably the most famous science fiction dynasty was Gene Rodenberry's *Star Trek*.

Star Trek

"SPACE: The final frontier. These are the voyages of the Starship *Enterprise*. Its five-year mission: To explore strange new worlds, to seek out new life and new civilizations, to boldly go where no man has gone before." These are the opening words to one of the longest-lived science fiction television franchises to date. *Star Trek* began in 1966 with a modest three-season production. Ordinarily that would signify a briefly successful television series, which would be quickly consigned to obscurity. But not *Star Trek*. The show's fans became known as Trekkies. After years of existing only in syndication, the franchise was rebooted in 1979 with a full-length film, followed by five more. The series was introduced to a new set of viewers in 1987 with *Star Trek: The Next Generation*, which took place about a century after the first show.

Trekkies now refer to the 1966 version of *Star Trek* as *Star Trek: The Original Series* (or TOS). The show that followed it, *Star Trek: The Next Generation,* is referred to as NextGen or TNG (1987–1994). In addition to these two series, there was also *Star Trek: Deep Space Nine* (DSN, 1993–1999) and *Star Trek: Voyager* (Voyager, 1995–2001). The plotlines of DSN and Voyager were contemporaneous with NextGen, and the story lines would occasionally cross. Finally, another prequel series called *Enterprise* (2001–2005) detailed the early years of mankind's experience with interstellar flight in the same universe. Combine this with twelve feature films, a cartoon series (1973–1974), hundreds of books, comic books, and other products, and you have a marketing behemoth.

With such a tremendous amount of material, there is no way it can all be described in a few pages. So the summaries below are just representatives of the Alien types and plotlines in the *Star Trek* universe.

The Original Series

The original *Star Trek* was born in the political tumult of the 1960s, a world in which racial integration, questions of gender equality, and Cold War proxy wars were dominant concerns of the American public. The series showed fresh alternatives to dealing with these problems. The starship *Enterprise* flew around the galaxy at speeds faster than light, captained by midwesterner James T. Kirk, but it also had a black female communications officer named Uhura, an Asian helmsman named Sulu, a Russian pilot named Chekov, a Scottish engineer named Scotty, a doctor from the American South named McCoy, and a first officer, Spock, who was an Earth-Vulcan hybrid. Vulcans were a warrior race that had tamed their violent tendencies through veneration and practice of logic. Vulcans rejected emotion as "illogical," and the show had recurring subplots with McCoy and Spock crossing verbal swords over the proper role of emotion. The crew of the *Enterprise* was ethnically mixed, highly functional, and happy. This was but the first implicit comment made by the show on the problems of the American 1960s.

The format was that the *Enterprise* would encounter some problem at the beginning of the episode that would be resolved at the end. Thus each episode stood more-or-less alone. The seventy-eight episodes covered many of the social issues of the 1960s. For example, the episode "Let That Be Your Last Battleground," described two Aliens, one black on the left side of his body and white on the right side, fighting another Alien from the same planet, this one black on the right side and white on the left. Their planet was destroyed from civil wars fought between these two very similar groups of people. In the end, the two Aliens headed back to their planet's surface to continue their battle to the death. The story line was an obvious reference to the black and white racial issues playing out in the United States at the time. Critics found the message to be a bit too obvious, but it was a common sort of plot device seen in TOS.

The *Enterprise* was the flagship of the political organization called the United Federation of Planets (usually called just the Federation), a consortium of planets and races that joined together voluntarily. Planets that had attained faster than light technology, were peaceful, and could adhere to democratic principles in their external dealings with other races were allowed to join, although there was no restriction on any planet's internal political and social organization. The main enemies of the Federation were the Klingon and Romulan empires. In TOS, the Klingons were very much human-like, although their features were what one might call vaguely Persian, with swarthy complexions and neatly trimmed heavy beards. The Klingons seemed to be paying an homage to *Flash Gordon's* Ming the Merciless. Americans might have heard of "Death before Dishonor" as a military motto, but Klingons exemplified the credo. The Romulans were also played by human actors and were consequently human in appearance, although with pointy ears and copper-based blood. They were wily and conniving. They were collectively depicted as a combination of earthly old Roman Empire ideals with a bit of classic Ming Dynasty thrown in.

The Next Generation

Taking place a century later than the original series, the starship was now the *Enterprise*, model D. Captain Jean-Luc Picard led a culturally diverse crew in a series of adventures. The crew included a blend of the Earth genders and races, along with some new Aliens among the leading characters. Deanna Troi was a hybrid between a human and a Betazoid, which was a species of human-looking telepaths. In addition, the century had brought about political changes, and the Klingons were no longer enemies but allies. The security officer on the *Enterprise D* was Worf, a Klingon who had been adopted and raised by a human couple. In the transition from TOS to NextGen, Klingons had been reimagined, and they were now portrayed by huge humans with makeup that added a set of bony ridges to their foreheads. The change was never explained in the show but was discussed in fan-generated fiction, as well as in books approved by the show's creators. The official explanation was given in DSN. It was an experiment in genetic engineering that went awry. The NextGen Klingons were the real ones. In NextGen, Klingons were better developed in the fictional sense. Their society was focused on honor and advancement through combat.

With Klingons now allies, new races of enemies were encountered, including the Cardassians, who played a central role in DSN, again humanoid, with a pronounced neck ridges and Orwellian culture; and the Borg, a culture of cyborgs. The Borg were not any specific race, as they incorporated all life-forms they came across. When they encountered any new species, they would broadcast: "We are the Borg. You will be assimilated. We will add your technological and genetic distinctiveness to our own. Resistance is futile." They were a powerful culture and the species they encountered were, indeed, often assimilated. However, we only encountered humanoid Borg due to the need to contain costs and given the technology available to the show's creators.

Another entity often encountered in NextGen was called Q. Originally apparently a single entity, we learned that the Q were actually a race of super powerful beings, with essentially godlike powers. Q could, in an instant, change reality, travel through time, destroy planets and stars, and kill or bring people back to life. It was never fully explained why Q would ever have any interest in the comparatively primitive Federation.

Other *Star Trek* Spinoffs

DSN, *Voyager*, and *Enterprise* introduced new races and political situations. The series were ongoing examples of how Aliens were no longer a novelty like they were in the 1950s movies, but rather were just characters to be used to advance the plot. In a sense, the various series were just a throwback to Homer's *Odyssey* or Jonathan Swift's *Gulliver's Travels*. The races of Aliens encountered were interesting, and their differences from humanity were often the basis for the story. However meeting an Alien was run of the mill, just another type of diversity to either embrace or recoil from, but in either instance an opportunity to learn.

Without a doubt, the *Star Trek* fandom—nicknamed Trekkies and calling themselves Trekkers—is the most well known in all of science fiction. As of this writing, the literary empire of *Star Trek* is 46 years old, with another movie released in 2013. The franchise is alive and well, and I hope it will continue to, in the words of Spock, "live long and prosper."

Star Wars

If *Star Trek* had a cerebral component, the *Star Wars* franchise was pure fun. *Star Wars* had a much different goal, which was to tell a classic adventure tale of a boy who didn't realize his princely origins, a princess in distress, and a powerful and evil adversary. *Star Wars* is a timeless story, in a setting "a long time ago, in a galaxy far, far, away."

The *Star Wars* franchise began as a single movie released in 1977. The story of the film included vital plans stolen from an evil empire. These plans are for the Death Star, a mobile battle station with sufficient power to blow up an entire planet. A princess by the name of Leia was bringing the plans to troops rebelling against the Empire. Before the princess is captured by the malevolent Darth Vader, the plans are placed in a robot (called a droid in the movie), and they make it into the hands of a farm boy named Luke Skywalker. He teams up with a Jedi knight called Obi-Wan Kenobi. The Jedi were once the protectors of the galaxy, part philosophers and part warriors, keeping the peace with their trademark weapon, the light saber. Obi-Wan and Luke discover the plans and resolve to bring them to the Rebels. They enlist the help of a smuggler named Hans Solo and his copilot called Chewbacca. Chewbacca is a Wookie, which is the first Alien main character we encounter, a 7-foot-tall hairy humanoid, reminiscent of Bigfoot and what one character called "a walking carpet."

The group escapes Luke's home planet, but their ship is captured and drawn into the Death Star by a tractor beam. They figure out a way to neutralize the tractor beam and escape. However, prior to the escape, Obi-Wan Kenobi engages in mortal combat with Darth Vader and loses. In addition, Luke's group realizes that Princess Leia is on the Death Star and free her.

The spaceship escapes the Death Star with the remaining occupants. They bring the plans of the Death Star to the Rebels, but a homing beacon that the Empire had surreptitiously placed on their spaceship brings the Death Star to the base to destroy it. An attack on the Death Star by small Rebel fighters is ultimately successful and the Death Star is destroyed.

The amazing thing about *Star Wars* is that the Aliens are not called out as being particularly alien. They are just simply characters. Chewbacca's "alienness" is never a concern. There is an iconic bar scene in which the patrons, entertainers, and workers are essentially all Aliens and they're just there for color. Alienness is essentially identical to our concept of race . . . something there, but little remarked upon, at least among the more cosmopolitan and liberal among us.

If the original movie could have easily stood on its own, its commercial success guaranteed that there would be a sequel. The second movie was called *The Empire Strikes Back* (1980), followed by *Return of the Jedi* (1983). These two movies were really one big movie split in half, and they described Luke's coming into his own and realizing that his father was Darth Vader. *The Empire Strikes Back* introduces Yoda, a Jedi master who has hidden from the Empire. Yoda is definitely an iconic Alien, known on sight to adults and children alike for more than 30 years. These three stories tell of the overthrow of the Emperor and the Empire, as good triumphed over evil. The story spawned hundreds of books that were very popular with young readers and avid science fiction fans.

The franchise received a reboot in 1999, when George Lucas decided to tell the origins of Darth Vader. The movie *The Phantom Menace* began with Luke's father, Anakin Skywalker, as a young boy and how he encountered the Jedi and subsequently became one of the most powerful Jedi knights. The two subsequent movies *Attack of the Clones* (2002) and *Revenge of the Sith* (2005) tell the tale of Anakin's mastery of Jedi powers and his slow corruption and change to evil, culminating in his becoming Darth Vader.

Many Aliens are encountered in the *Star Wars* story arc. As we have said, Yoda is a short green Alien, while Chewbacca is essentially Bigfoot. We meet Jabba the Hutt, a member of a large slug-like species. The Hutt are essentially gangsters, although it is not clear if this is a characteristic of the species or just the family we encounter. The Gungans are an amphibious species.

In *The Phantom Menace*, we meet Aliens who are obvious stand-ins for common earthly stereotypes. These stereotypes are not the politest and most admirable and the filmmakers have rejected any claims that they intended to incorporate stereotypes. Still, the Gungan have been criticized as being recognizable caricatures of African-heritage Caribbeans. There are also caricatures of Asians of the sort commonly seen in World War II movies—a sort of Fu Manchu—in the character of Nute Gunray, a Neimoidian. The character Watto is a Toydarian owner of a secondhand store and has likewise been criticized as a caricature of an Arabic or Jewish shopkeeper. (Personally, my impression was not of a Semitic individual but rather of a classic movie-depiction of a shady, nonmobster, recent immigrant Italian.) It is possible to accept the filmmaker's protest of innocence and instead simply see cowardly or greedy or bumbling characters who have been historically portrayed in movies with specific, recurring characteristics.

We can go so far as to say that the fact that Aliens are associated with Earth stereotypes at all is a testament to their evolution in the minds of the public. It actually highlights just how nonalien the Aliens of *Star Wars* really are. We readily identify their forms as alien, but their behavior is actually quite familiar to moviegoers, and we can recognize the classic Hollywood stereotypes, even if they don't have the right visual cues. This says a lot about the degree to which Aliens have been accepted in our culture and the degree to which the *Star Wars* filmmakers have incorporated commonly used Hollywood caricatures. This movie franchise has amassed over two *billion* dollars (four billion when adjusted for inflation), suggesting that a movie in which Aliens are accepted as "just folks" will be commercially successful and can perhaps shape the public's viewpoint that Aliens will be much like us. This is likely to be factually incorrect, but that's not the story that Hollywood is telling us.

Alien

If the *Star Wars* franchise presents Aliens that are recognizably human, the movie *Alien* depicts something very different. The Alien in *Alien* isn't named, but it is called a xenomorph. It is unclear if the Alien is intelligent in the commonly used meaning of the word. It is a eusocial species, essentially structured on the idea of a wasp colony. Their "society," for lack of a better word, consists of a single, egg-laying queen, along with a caste of warriors. The movie *Alien* was released in 1979, with three subsequent sequels: *Aliens* (1986), *Alien 3* (1992), and *Alien Resurrection* (1997). Two additional movies bridged the Alien universe and another series called *Predator*.

The *Alien* movies have only the most modest of plots. The xenomorphs are supreme hunters, and humans are simply food or organisms that facilitate xenomorph reproduction. The *Alien* queen lays eggs, which contain a crab-like form of the alien that grabs a human's face and implants an embryo through the mouth. The face-hugging form falls off, and the embryo grows. Eventually the embryo burrows out of the human's chest, killing it. The xenomorph grows into its large form, which captures humans and brings them to back to the eggs to gestate additional Aliens (figure 4.1).

That unarmed humans are no match for the xenomorphs is a central theme of all of the movies. The entire franchise is an extended nightmare, with tremendous carnage. A single female human protagonist named Ripley is the heroine who we root for throughout the series, and she successfully

FIGURE 4.1. The Alien in the movie *Alien* is humanoid because filmmakers of that era did not have access to today's computer graphics technology and therefore the Alien had to be played by a human actor. However, the behavior of this Alien is not at all human-like. Instead it is a ravenous monster from mankind's dark psyche. 20th Century Fox.

defeats the enemy in each movie. While the *Alien* franchise is clearly a movie about Aliens, it is really just a horror film and taps into the same sort of fear as the movie *Jaws*. Movies in which humans are helpless victims are just plain scary.

Stargate

The *Stargate* juggernaut consists of both a movie and an extensive television franchise, with a couple additional movies that were released only on DVD. On television, shows that were inspired by the original movie ran for about 14 years, with three distinct series.

The premise of the original movie (1994) is that archaeologists find a large "Stargate" in Giza in 1928. The Stargate consists of a large hoop with glyphs running around the perimeter. It takes until 1994 before it is understood that this hoop is actually a bit of Alien technology. It has been moved to a U.S. military base located at Creek Mountain and powered up. Nobody has any idea what it really does until an archaeologist deciphers the glyphs and realizes that the hoop must be turned in a specific sequence like the combination lock of a safe. When done properly, the Stargate is opened, connecting our world to another Stargate on another world.

A military team is sent through the gate with the archaeologist, and they encounter a world that is much like ancient Egypt, desert and sandy. They find a powerful Alien who once lived on Earth as the Egyptian god Ra. He has enslaved humans to serve him, and it becomes clear that the culture of ancient Egypt began because of a visit by this immortal Alien. The military team is able to initiate a rebellion among the slaves and Ra attempts to escape in a spaceship that is easily recognizable as an Egyptian pyramid. However, the military team had brought with them a tactical nuclear weapon to detonate if they encountered a situation that was deemed dangerous to Earth. They were able to put the weapon on Ra's spacecraft, and it exploded as the spacecraft was leaving the planet's atmosphere. The military team returned to Earth through the Stargate, leaving behind the archaeologist.

The idea that Egyptian culture was influenced by ancient Aliens is a familiar one, made popular by Erich von Däniken. It is unclear how successful this franchise would have been, had the public not been primed by such books as *Chariots of the Gods*.

The original movie was commercially successful, bringing in about $200 million. This led to the television shows. *Stargate SG-1* (1997–2007) had 214 episodes, *Stargate Atlantis* (2004–2009) had 100, and *Stargate Universe* (2009–2011) had 40. In the various shows, the creators had ample time to expand their vision of the *Stargate* universe and the inhabitants found within it. The plot line was somewhat like *Star Trek* in that the episode of the week involved going somewhere, encountering some problem, and solving it. In doing so, the protagonists encountered a dizzying panoply of Aliens. It is quite impossible to cover all of the species and their various interconnections. Let's look at two races closely, as they involve archetypes we have encountered before.

The Goa'uld are Aliens of the type found in the original movie. It turns out that Ra, who was a somewhat androgynous human, was merely a human host for the real Alien. The Goa'uld are snakelike parasites that can attach themselves to a brainstem and control the human host. This symbiotic connection confers long life on the carrier, at the expense of losing the carrier's personality. The Goa'uld are compassionless and are bent on galactic domination. They

FIGURE 4.2. The Asgard of *Stargate* have many of the characteristics of stereotypical Alien Grays. MGM Television Worldwide Productions.

are noteworthy in our context, as this particular variant of Alien was supposedly on Earth in our history, as suggested by von Däniken (although von Däniken was very light on the details). According to the show's mythology, humans would not have been able to build the pyramids without the technology of the Goa'uld.

Another important Alien from the *Stargate* universe are the Asgard. The Asgard are a declining species who have lost their ability for sexual reproduction and must clone themselves to survive. Although in the distant past they were human in appearance, the repeated cloning has led to degradation in their form. The Asgard now are the canonical "Gray" Aliens of the type seen by Betty and Barney Hill (figure 4.2). The Asgard visited Earth in the prehistory of early Nordic tribes, including the Vikings. In fact, one of the main Asgard characters in the series is Thor, the original Norse god of thunder. The Asgard are technologically advanced and fight the Goa'uld. The Asgard are also under attack by another Alien species called the Replicators, a race of robotic Aliens that attack and assimilate other species' technology. The Replicators are somewhat like the Borg of *Star Trek*.

Many, many Aliens are encountered in the *Stargate* universe, with the bulk of the story lines having to do with how they shaped Earth's history. In a later multiyear story arc of the series, an Alien species is found to be responsible for the legend of Merlin in King Arthur's saga. The Aliens of *Stargate* have motivations that are generally recognizable as being pretty human. Conquest, aggression, strife, and defense—we can relate to these characters because they are motivated very much like we are.

X-Files

We mentioned the television show *The X-Files* at the beginning of chapter 2, and we return to it here. The examples of Aliens represented in the movies and television shows we've been talking about up until now were obviously fictional, and there was no real attempt to make them sound believable. However, there *are* movies and television shows that depict fictional Aliens much more closely to what Americans believe that Aliens "really look like." We'll talk about two of them here. The first one we'll discuss is *The X-Files*.

The X-Files premiered in September 1993 and ran for nine seasons through 2002. During its heyday, it was the longest-running single science fiction series in American television history, although it was subsequently passed by *Stargate SG-1* in 2007. *The X-Files* resonated with a country that had gone through Watergate and Iran-Contra and was highly suspicious of the government and what it was doing behind the scenes. In addition, *The X-Files* was born in the same environment that brought us the fake Roswell Alien autopsy, also shown on the Fox Network.

That *The X-Files* tapped into the American distrust of government can be seen in some of the taglines used to advertise the show: "Trust No One," "I Want to Believe," and "The Truth Is Out There" (figure 4.3). The show followed two FBI agents (Fox Mulder and Dana Scully) as they investigated cases that were deemed interesting but unsolvable. As an example, according to the mythology of the series, the first X-File was started in 1946 by FBI Director J. Edgar Hoover. The file contained information about a series of murders that occurred in Northwest America during World War II. The victims were killed, ripped to shreds, and eaten. It looked like they had been attacked by a large animal, but the victims were found in their own homes with no signs of forced entry, and it appeared as if they had invited the killer to enter. Agents investigating the murders cornered what they thought to be the animal in a cabin and killed it. However, when they entered the house, they found no animal, but rather the dead body of a man named Richard Watkins. Hoover deemed the case unsolved and filed it away. While fictional, the tale is reminiscent of a werewolf story. *The X-Files* has some similarities to the far less successful 1974

FIGURE 4.3. This advertisement for the television series *X-Files* tapped into the undercurrent in American society that believes that the Earth has been visited by Aliens and that the government knows more than it is telling us. 20th Century Fox.

television show *Kolchak: The Night Stalker*. The original X-File didn't involve extraterrestrials, but they appeared soon enough.

The X-Files had two basic plots, the first was "monster of the week," in which cases involving werewolves, vampires, and other supernatural creatures were investigated. These shows were pretty much stand-alone plot lines. However, throughout the show, there was also the recurring story of Aliens, alien visitation, and the idea that the government knew more than it was telling us. The paranoia of Roswell was well represented in this television series.

In the show, Mulder was a firm believer in the supernatural and extraterrestrials, while Scully was the skeptical scientist. In many shows, Scully was able to rationalize the evidence they uncovered but never completely. Over the course of the series, Scully became increasingly dissatisfied by her inability to explain away what they found. She never became the believer that Mulder was, but she was more willing to give greater weight to his far-ranging theories.

Mulder's belief in Aliens stemmed from his sister's having been abducted when he was 12 years old. This incident threads itself throughout the series. Mulder and Scully become allies as they encountered a sinister and shadowy arm of the government called the Syndicate. The Syndicate is comprised of the classic "Men in Black," agents who cover up inconvenient incidents that the government doesn't want the public to know about. Mulder and Scully's primary antagonist is an agent known only as the "Cigarette-Smoking Man." He is a ruthless and compassionless killer and a man with powerful connections. It could be said that he has the Illuminati on speed dial.

The Syndicate is mankind's liaison with Aliens intent on taking over the Earth. These shadowy figures have not only infiltrated the American government but all governments. They are the "real" power in the world. This theme explains the series popularity with the conspiracy theorists among us.

The series ends with Mulder's being subjected to a secret military tribunal, charged with breaking into a top secret military base and viewing the plans for the Alien invasion and subjugation of the Earth. Mulder is found guilty but escapes with the help of other agents, and he and Scully become fugitives.

The impact of the show *The X-Files* is hard to judge. On the one hand, it is fiction; on the other hand, it reinforces ideas present in our society. There are many among us who claim that there is more going on than we are led to believe. Conspiracies like those said to be related to the JFK assassination, the destruction of the Twin Towers on 9/11, the Illuminati, and so on, are believed by some and suspected by others. We know of incidents when the government withholds or manipulates information to lead us the way it wants to. The show's tagline "Trust No One" reinforces our collective paranoia.

As a scientist, I am concerned about an even more pernicious effect. In *The X-Files*, the two main characters represent the believer and the skeptic. As the show progresses, the skeptic becomes less skeptical, showing that the believer was right all along. In the show, the characters encounter data that makes this a reasonable progression, but it is, after all, fiction. However, I worry that shows of this type reinforce the dangerous idea that irrational people are more rational than the rational. It's one thing to have an open mind. It's quite another to think that a literal werewolf is a real possibility.

Still, as a piece of fiction, *The X-Files* is an excellent show, which reflects a particular vision of Aliens that is commonly found in our culture. The saga of Betty and Barney Hill and Roswell is alive and well.

Close Encounters of the Third Kind

Close Encounters of the Third Kind (1977) is one of the better Alien movies from the point of view of incorporating "real" phenomena as reported by people who claim to have had some extraterrestrial contact. Because the movie touches on so many of the classic extraterrestrial stories, I describe the movie in more detail than I have in other cases and point out the depiction of iconic extraterrestrial elements as they are occur in the narrative.

A little, glowing, red will-o'-the-wisp depicts the foo fighters of 1945. Multiple, multicolored glowing craft, zooming through the air, are classic UFOs, which include observation by passenger aircraft in-flight. People who encounter the flying lights feel compelled to go to a certain place at a certain time (as was reported to have occurred with George Adamski). People were abducted, and the government knows more than it's telling. One element of Alien lore that we did not discuss in chapter 2 occurs: military pilots who "just disappeared" while on a flight mission are returned. When finally observed in person, the Aliens turn out to be Betty and Barney Hill's Grays. The movie is a colorful spectacle, reflecting an artistic and masterful directorial eye.

The story opens in the present day (1977) in the Mexican Sonora desert, where Mexican soldiers or police have discovered a circle of vintage World War II fighter planes in perfect condition. An American team shows up and looks over the situation and finds that the planes start easily. By looking at the serial numbers on the engine blocks, they are identified as being "Flight 19," which was lost off the Florida coast in 1945. (Fact: Flight 19 took off from Fort Lauderdale Naval Air Station on December 5, 1945, with five Grumman Avenger Bombers. All planes were lost, including a float plane that went out to look for them. UFO enthusiasts have long noted that the planes disappeared in the Bermuda Triangle.)

While in the desert, we encounter a French-speaking UFO expert (although we don't know this yet) and his interpreter, who was a cartographer by trade. They find an old man who tells them that the planes had just appeared in the middle of the night. He is sunburned and he tells them, "The Sun came out last night and it sang to me."

In a series of little vignettes, the director shows us an encounter between a UFO and a TWA flight outside Indianapolis. In a rural house in nearby Muncie, a little boy named Barry is asleep, when his electrically powered toys go crazy. Awakened, he sees flashing lights outside and, when he goes downstairs, he sees that the house's appliances are all on as well. He heads outside to look at the lights. By now, the blinking and moving toys awaken his mom, Jillian, who goes out into the night to search for her son.

In the next scene, we meet our main character, Roy, who works for the power company in the Indianapolis area. He gets a phone call that the electricity is out all over the place. Roy is sent out to a specific area to see if he can get the equipment back online. First he sees a bunch of mailboxes jumping around and a railroad crossing sign shaking wildly. Then the batteries die in his truck and flashlight. A bright light shines down on the truck, giving him a sunburn over the half of his face that was looking up. The light goes out, and he sees a darkened, slow-moving UFO flying overhead.

Roy hears radio reports of UFOs in another part of the county and speeds over there, nearly hitting little Barry, who is rescued in the nick of time by his mom. While Roy is making sure everyone is OK, the wind picks up and several very colorful UFOs fly over, including a little floating red light the size of a basketball. This light is highly reminiscent of the foo fighters of World War II.

Roy goes home and drags his wife and kids back out to look for more UFOs. They see none, and his wife starts to suspect his sanity. The next day, the newspaper's headline reads, "UFOs Seen over Five Counties," but his wife hides the story from him. Roy begins to be fascinated by mounds of stuff: shaving cream, a pillow, a mud pile, mashed potatoes. He is drawn to try to sculpt them but without really understanding why.

The film then changes direction to a scene in India, in which a large crowd is sitting and repeatedly chanting a particular five-note tone that they had heard the night before coming from the sky. The French UFO expert and his translator are there to record the chant. The two characters are next seen at a conference and then a radio telescope facility that has been receiving music tones similar to the chant in India. The telescope is also receiving a series of numbers, which the cartographer deciphers as latitude and longitude coordinates. The location: the Devil's Tower in Wyoming.

We jump back to Jillian and Barry. Barry is playing the same five-tone chant on a toy xylophone, and Jillian is drawing mountains. Outside, it looks like a storm is coming, with roiling clouds. In a scary sequence, UFOs surround the house and Barry is abducted.

The American UFO experts, including the Frenchman and his interpreter, have initiated an expedition to Devil's Tower. They devise a way of evacuating the area and settle on a fictional deadly chemical spill. We see a military unit getting together the required supplies and shipping them to Wyoming in trucks disguised as standard commercial transports.

Roy, meanwhile, is trying to bring some normalcy to his life, but his behavior still troubles his wife. While making yet another mound, he becomes frustrated and pulls off the top, leaving a mesa. Suddenly, he knows what he was missing. In an episode of what could easily be considered craziness, he goes outside and yanks plants and bushes out of the yard and throws them into the house through his kitchen window. He tosses spades full of dirt, a garbage can, and chicken wire. Overcome by the spectacle, his wife takes the kids and leaves. Roy makes in his living room an 8 feet tall detailed model of a mesa.

The stories start to come together when the government reports the fake chemical leak on the television. Roy sees it and realizes the mesa in his living room is obviously a model of Devil's Tower. Meanwhile, Jillian sees the same news coverage, and we see that she has been drawing the mesa herself. Roy and Jillian independently realize that they need to travel to Wyoming.

When Roy gets to Wyoming, he stumbles on what might be called a circus atmosphere of evacuation. He meets Jillian and, since they don't know that the chemical spill story is a fake, they buy gas masks from impromptu vendors in the circus. They set off toward the mesa in his station wagon, bypassing army checkpoints and heading to Devil's Tower. Unfortunately, they run into an army convoy and are captured. They are brought to an assembly point, where the French UFO expert is stationed. Roy gets interrogated. There is an argument between the UFO expert and the army commander, and the army wins. Roy and Jillian are put on a helicopter (along with many other people with similar stories) to be evacuated. At the last moment, Roy and Jillian make a break for it and head up the mesa. As they crest a rise, they see a base on the other side.

The base looks a little like a big helipad, surrounded by cameras and klieg lights. A voice on the intercom tells them that there are radar contacts and the base bustles as scores of experts take their position. The familiar UFOs and foo fighters arrive. As they hover over the helipad, the humans play the five-note tone. The notes correspond to lights on a large display behind them. After a few attempts, the UFOs play the tones back, but then the UFOs leave.

While the experts begin congratulating themselves, a huge roiling of clouds appears on the other side of the mesa, heralding the arrival of the mother ship (figure 4.4). What follows is what one might call a high-tech version of "dueling banjos." The humans and Alien craft play passage after musical passage, copying one another, all the while accompanied by synchronized colorful lights.

At the conclusion of the lightshow, the bottom of the mother ship opens, and many humans disembark, including World War II pilots, perhaps from Flight 19. None of the people have aged at all. Barry also leaves and is reunited with his mother. The door of the ship closes and reopens, this time with Aliens emerging. The first to emerge is a tall and very thin version of the familiar Grays, followed by dozens of more traditional Grays, perhaps 4 feet high (figure 4.4).

FIGURE 4.4. The mother ship in *Close Encounters of the Third Kind* (*left*) was huge, many hundreds of feet across, while the Aliens (*right*) were classic examples of the diminutive Grays. Two humans are shown for scale. The black-and-white figure showing the mother ship doesn't do it justice. You really have to see the movie to appreciate the magnitude of the spectacle. Columbia Pictures Corporation.

The humans had prepared a group of astronauts who they hope will leave with the Aliens. At the last minute, it is decided that Roy will join them. Roy is welcomed by the Grays with open arms, and he is guided onto the ship. It is implied, but not obvious, that he is joined by the other astronauts. The Aliens all reenter the ship, and the door closes for a final time. As it rises majestically into the sky, Barry closes the movie by saying, "Bye!"

Close Encounters of the Third Kind incorporates many of the "right" elements as believed by UFO enthusiasts, and it resonated well with that community, although, as always, there were purists who quibbled with this point or that. The movie was a huge commercial success, grossing more than $300 million worldwide.

The title came from a scale devised by astronomer and UFO researcher J. Allen Hynek and popularized in his 1972 book *The UFO Experience: A Scientific Inquiry*. His scale classified UFO sightings as close encounters of the first kind and sightings with physical evidence, like scorch marks or lost time (à la Betty and Barney Hill), to be the second kind. Close encounters of the third kind required that one encounter "animate beings" with the UFO. The name was vaguely chosen to allow for the possibility that perhaps UFOs were not extraterrestrial in origin. There have been subsequent extensions of the Hynek scale, but these are not universally accepted. Fourth is abduction with retained memory. Fifth is for regular conversations (like the Adamski experience). Sixth is an encounter that causes injury or death to a human.

Finally, close encounters of the seventh kind requires human/extraterrestrial mating that produces an offspring, often called a "star child."

The idea of this interspecies mating has been reported by some of the post–Betty and Barney Hill abductees and also proposed by von Däniken and his contemporaries as possible explanation of, for example, the human/beast hybrid gods of ancient Egypt. Even a cursory knowledge of genetics shows how ludicrous this idea is. Think about it: humans and oranges share a genetic history and have appreciable genetic overlap, yet a human/orange hybrid is unthinkable. In contrast, Aliens and humans have no shared genetic history; indeed it is unlikely that the genetic material of Aliens looks much like the DNA of Earth-based life. From mankind's recent advances in genetics, we know that it is possible for genetic material from one species to be transplanted to another, but the merging of human and Alien genetic material seems very unlikely.

Nonserious Depictions of Aliens

While we have spent a lot of time talking about Aliens as they are depicted in literature, radio, movies, and television, there is also a class of Aliens that we aren't meant to take seriously. These are just representations that allow us to watch a charming or funny story.

Of the nonserious depictions of Aliens, perhaps the most serious is the 1982 movie *E.T. the Extra-Terrestrial*. A group of Alien botanists sampling the plant life on Earth are startled and they take off, leaving one of their members behind. The Alien encounters a ten-year-old boy who helps him get back to his own kind. The movie does utilize some of the classic Hollywood techniques to tell the story of Alien contact, for instance when the government gets involved and tries to capture E.T. for study. But, given its intended audience, it is difficult for the movie to seriously depict Aliens.

In television, Aliens are sometimes used as the basis for a sitcom. The silly behavior of creatures who have no idea how human society works can easily be exploited for laughs. *Mork and Mindy* (1978–1982) used the frantic humor of Robin Williams as Alien Mork tries to live among us. The eponymous *ALF* (for Alien Life Form, 1986) was played by a hand puppet. He was born on the Lower East Side of the planet Melmac. He crash-landed in California and moved in with a family he encountered. He is a sarcastic wise guy and is often trying to eat the family's cat. *My Favorite Martian* was a 1963 television show with a similar premise. A Martian anthropologist crash-landed on Earth and moved in with a human while he tried to repair his craft. Only his roommate knows he's a Martian. To a degree, there is a similarity to the more famous *Bewitched*, in which Samantha the witch lives among us, known only to her husband.

One of the most iconic of the nonserious Aliens is Marvin the Martian. Marvin debuted in the 1948 cartoon *Haredevil Hare* opposite Bugs Bunny. He is dressed as a Roman centurion, in homage to the identification of the planet Mars with the Roman god of war. Marvin is an astronomer frequently bent on destroying the Earth "because it obstructs [his] view of Venus." The way he will destroy the Earth is with "an Illudium Pu-36 Space Modulator" (sometimes "Illudium Q-36 Space Modulator"). As with any character that has a conflict with Bugs Bunny, Marvin has very little luck with his plans. A common phrase when he fails to blow up the Earth is, "Where's the kaboom? There was supposed to be an earth-shattering kaboom!" Marvin the Martian cartoons are often quite funny, but he is never intended to be taken as representing a real Alien.

Another Alien who is somehow not an Alien at all is Superman. Superman, born Kal-El of the planet Krypton, was sent to Earth before his planet exploded. By virtue of having been born on the high-gravity planet Krypton, Superman is very strong. "Faster than a speeding bullet, stronger than a locomotive, and able to leap tall buildings in a single bound" was a literal description of his abilities, but these morphed over the years until Superman was indestructible. Superman is not an iconic Alien but rather a superhero.

The 1997 movie *Men in Black* taps into a little of the lore that surrounds "real" Aliens. A super-secret government organization of men in black suits regulates the comings and goings of Aliens living here on Earth. People who encounter the many Aliens among us are subjected to a "neuralizer," a little gizmo that flashes a light and causes people to forget. In this movie there are dozens of different kinds of Aliens, from a bunch of partying cockroaches to a huge worm the size of a New York subway train. The movie, along with its sequels, is great fun. Aliens in movies are now commonplace. Rather than being alien, they are just part of the plot, like the inexperienced and grizzled partners in some cop/buddy movies, or the hopelessly mismatched couple in many romantic comedies. Aliens have evolved to the point where they are no longer a novelty.

Aliens Not Mentioned

In a genre as rich as science fiction, it is inevitable that some readers will object to Alien depictions not mentioned here. The 2009 movie *Avatar* was a blockbuster success and a gorgeous movie, but it is too soon to see if the Na'vi will become an iconic fictional species. *Doctor Who* is a long-running series describing the antics of a "Time Lord," a member of an alien species with the ability to travel through time. While wildly popular among a growing cult following, *Doctor Who* hasn't made it into broad awareness of the general public outside the United Kingdom. The inability of a particular science fiction story to transition into common knowledge is the most frequent reason for why that particular story was not described here in detail. *Transformers, Predator, Independence Day, Third Rock from the Sun, The Coneheads, V, Battlestar Galactica, Starship Troopers, Blade Runner, The Hitchhiker's Guide to the Galaxy, The Fifth Element, Dune, Firefly, Lost in Space*; the list can go on for a long time. Similarly, there are brilliant authors of pulp (and contemporary) fiction that haven't been mentioned: Fredrick Pohl's *Heechee*, Larry Niven's *Tales of the Known Universe*, Robert Heinlein's *Stranger in a Strange Land,* along with his delightful body of work, Ray Bradbury's *The Martian Chronicles*, Larry Niven and Jerry Pournelle's *The Mote in God's Eye*, this list is also long.

I have also not mentioned Aliens from video games. From the original *Space Invaders* to the bad guys in *Starcraft, Quake, Halo*, and so on, the problem with video games is that the Aliens tend to be known by a small and enthusiastic band of gamers. It may yet happen that a video game Alien species will become generally known to the general public, but this has not yet happened.

So let me apologize to all readers for their favorite Alien I haven't mentioned. I love them all too.

Alien Archetypes

Now that we've talked about the history of Aliens and how we met them, we're ready to summarize the archetypical Aliens. This repeats an earlier exercise, but now includes creatures we encounter in both fiction and "true" Alien stories. There are many different types, and we now know their origins. Most of these are encountered in fiction, and only a few are significant players in the "real Alien" mythos.

Little green men. These appeared more in the pulp fiction era and were precursors of Grays. They are found in some of the 1950s UFO movies, which, although often black and white, somehow convey a green-ness to their Aliens. Green Aliens are still found in children's movies, like *Toy Story*.

The Grays. These are the most common form of Aliens in reported encounters and in any movie in which the Aliens are nominally real. They are called

"Grays" because of the color of their skin. They typically have a large head and forehead, small chin, no nose and almond-shaped, black eyes. The origins of this variant of Alien appears to be the Betty and Barney Hill incident. The eponymous *Paul* is an alien of this type, as are the Asgard from *Stargate*, the Visitors from the book *Communion*, and the Aliens from *Close Encounters of the Third Kind*.

Angelic Space Brothers. This type of Alien was first encountered by George Adamski. They vary somewhat, but they are described as tall, beautiful, and Nordic in appearance, generally with long hair. (Adamski's Alien was actually relatively short.) These Aliens are very spiritual and have come to teach us about cosmic harmony. They tend to be a bit arrogant and their motivation for contacting us is to save us from self-destructive behavior. Occasionally they warn us to improve how we act or they will somehow keep us on Earth until we do. Some Alien skeptics have noted that this variant of Alien is very similar to the role that angels once played in society when religion was accepted more universally and that they really have the same function, which is providing an object lesson in knowing how we should behave. Those who believe that the Space Brothers are real point to our legends of angels as proof that they have visited the Earth in the past. One example of this type of Alien is Klaatu from *The Day the Earth Stood Still*.

Evil insects. They vary in their range of intelligence, so it is ambiguous whether they count as Aliens or mere alien life-forms. Accordingly, whether they are evil or not depends on their intent. Typically they are hunters and killers of humans. The Aliens in the movie *Alien* and its sequels have an ambiguous intelligence. Perhaps they hunt just to eat and reproduce. The Aliens from *Starship Troopers* appear to have a sort of hive mentality, with some of the aliens encountered being swarming fighters, while others are more intelligent. We can also include the Formies from *Ender's Game* here.

Warriors. These are Aliens who value honor, bravery in battle, and aggressiveness above all. They consider life-forms who do not yearn for combat to be weak and therefore creatures to be conquered and either exterminated or enslaved. The Klingons from the *Star Trek* universe are iconic versions of this type of Alien, especially those from *Star Trek: The Next Generation* onward. The Green Martians of Edgar Rice Burroughs's Barsoom universe are other excellent examples. The Hawkmen of the Flash Gordon comics are warriors, as are the Kzinti of the Larry Niven *Ringworld* universe. It is possible that the eponymous *Predator* Alien counts, although it is not completely clear whether his race are hunters or warriors out for a little rest and recreation. Another variant of this archetype are Aliens who are a fighting class of a larger society. Often this variant is not the leader class of the society, and the Jem'Hadar of *Star Trek: Deep Space Nine* and Jaffa of *Stargate* typify this type.

Cuties. These Aliens are usually designed to get our children to have us part with our hard-earned money. They are cute, often evocative of warm and fuzzy pets, teddy bears, or other cuddly memories. The Ewoks of *Return of the Jedi*, the Tribbles of *Star Trek*, E.T. of *E.T.: The Extraterrestrial*, and maybe ALF are cutie aliens.

Yankee traders. While the historical and earthly Yankee traders were interested in making money, the Alien variant ranges from the merely acquisitive to species for which money is central to their culture. The Ferengi of the *Star Trek* universe are one example, as are the Psychlos of L. Ron Hubbard's classic pulp novel *Battlefield Earth*.

Shape shifters. These Aliens have an unspecified natural form but can assume the shape of others to blend in, sometimes to hunt. Examples include the changeling assassin in *Star Wars: Attack of the Clones*, the unnamed Alien in John Campbell's *Who Goes There*, the pod people of *Invasion of the Body Snatchers,* or the race personified by Odo in *Star Trek: Deep Space Nine*.

Mechanical organic life haters. These are a mechanical form of life or occasionally a mix of organic and robotic components. More often than not, they are driven to exterminate or enslave organic beings. The Borg of the *Star Trek* universe are one example, as are the Cylons of *Battlestar Galactica*. *Star Trek: The Original Series* exploited this form of Alien frequently, with the episodes "The Doomsday Machine" and "The Changeling," as well as V'ger from the first *Star Trek* movie. Doctor Who fans will recognize the Daleks as one of this type of Alien. A rare variant is the good robot, for instance the Autobots in the *Transformers* cartoons and movies.

Gods. These Aliens are so powerful that they can do anything. They often are capricious, sometimes malicious, and sometimes ambivalent. The Organians and the Q of the *Star Trek* universe are two examples, as are the Goa'uld of the *Stargate* universe.

Wrap Up and Transition

Thus far, we have discussed the Aliens we have imagined and even dreamed of. Because these Aliens have been our own creation (or have supposedly done us the courtesy of visiting us so we know what they look like), we have had some power over who and what they are; indeed they often are mirrors, reflecting our collective psyche. However there is a real question. Are there *real* Aliens in the universe? If we ever decide to leave the solar system and travel to nearby stars, what will we find? Are we alone or will we one day join a cosmopolitan galaxy, just one more species among many?

INTERLUDE

The inimitable Mark Twain once wrote in *Pudd'nhead Wilson's New Calendar* that "truth is stranger than fiction." Nowhere is that truer than in the discussion of extraterrestrial life. Thus far in this book, we have spoken of fiction and of stories that cannot be confirmed. While it is possible that one or more of the tales told by Arnold or Adamski or the Hills are a factual and accurate rendition of their experiences, anecdotes are an unreliable source of knowledge, no matter how gripping and entertaining.

For a question such as whether Aliens actually exist or what they might look like, we need to turn to science and, to paraphrase Twain, what we learn here is far, far stranger than fiction. Aliens are highly unlikely to be humanoid. The odds of them being able to eat us are nil. The range of the possible is so much broader than the strictures imposed by filmmakers and the need to have a recognizable plot line.

In the next pages, we will go in a different direction, one that is far more likely to teach us something about actual, physical Aliens, rather than Aliens as an earthly, social phenomenon. Biologists have explored some of the myriad possibilities of body plans seen here on Earth in the various mammals, birds, reptiles, and insects. More recent scientific research has considerably broadened our understanding of the various biochemical reactions that can lead to life. Breathing oxygen and exhaling carbon dioxide is an excellent way to keep an organism alive, but it is by no means the only way. Types of life on Earth can exist in environments that would kill you and me, but the range of possible environments on Earth pales compared with those on other planets, environments in which no earthly life would survive. However, scientists know of ways in which other chemicals can combine that would serve the same purposes as our familiar respiration and metabolism, some of which proceed at pressures that would compress you to the size of a pea and temperatures that freeze air completely solid. In order to understand the range of what life might look like, we must explore the range of the possible and delve into the restrictions imposed on life by the physical and chemical rules of matter itself. In these next few chapters, well explore what sorts of things govern the form of *real* Aliens.

It is important to remember that just because something is physically possible, doesn't mean it really occurs. If physics and chemistry allow for a particular kind of Alien, it could exist in a distant galaxy and we'll likely never encounter it. Thus, when we're asking about what Aliens we'll encounter if we venture into the galaxy, we should ask the simple question: "But what Aliens actually exist in our stellar neighborhood (if any)?" The safest way to do that is to simply ask them. Literally, as you read this right now, scientists across the globe are listening to the radio hiss of the heavens, hoping to identify the faint crackle that brings to us the voice of our neighbors. We'll talk about these scientists and their multidecade quest as well. So let's sit back and dive into what science teaches us about Aliens.

FIVE

LIFE-FORMS

You can see a lot, just by looking.

Yogi Berra

For the first half of the book, we've been discussing the history of mankind's vision of Aliens. If it is easy to see how our ancestors might have been interested in the subject, it is just as easy to see why the idea continues to fascinate us. The question of whether we are alone in the universe is one of the most compelling mysteries of all. This second half of the book explores our modern and scientific thinking. If we ever do meet an extraterrestrial, what is it likely to look like? Can we empirically explore the possibilities?

If we're going to talk seriously about Aliens, perhaps the place to start is to visit them in their home. Let me transport you to a world never seen before by human eyes. Go ahead and look around. Meanwhile, let me play tour guide and tell the other readers what you're seeing.

There are no trees in this alien land. There are plants, or at least things that look like plants, but they're unlike anything you've ever seen. Off to your left, a grove of unusual emerald fronds sway gently, rising high above you, like dozens of green ribbons might look if they were stirred by a breeze. Occasional rustling hints of something possibly moving through them unseen.

Those are the most familiar looking of the plants. Off to the right, a peculiar growth has a passing resemblance to a carrot, shorn of its greenery and balanced precariously, with nothing more than the skinny tip stuck in the ground. Only the shape is carrot-like, as the coloring and texture look like a pale strawberry and the cluster of spikes guarantee no bunny is ever going to make a meal of them. Other plants are weirder still. One looks like a cactus, except it is giraffe-spotted and topped by what could be tentacles, seven waving appendages that might or might not be dangerous.

The plant life is unfamiliar, but the animal life is downright freaky. The mystery of the shaking of the green ribbon plants is solved as a truly bizarre creature pokes its nose out from the undergrowth. Of course, "nose" is just a bias of your earthly experiences. As the creature emerges even further, its true shape is revealed. Maybe 5 or 6 inches long, the animal looks like a fat worm, walking on seven pairs of long, unbending, legs, like a Chinese dragon on many stilts. Sprouting from its back are fourteen long and dangerous-looking spines, a clear sign that something thinks of this animal as lunch.

Closer to you, the ground is covered with clean, white sand. A small chitinous creature scuttles around your feet perhaps grazing or possibly just out for a stroll. It looks like a horseshoe crab without a tail, or maybe just a huge beetle, with a segmented back and lots of feet. After it noses at your toes for a moment or two, it resumes its erratic journey. Your eyes follow it as it meanders away.

The sunlight is familiar at least. The bright yellow-white light shines from a clear blue sky, unmarred by clouds. A shadow flits over you, once, twice, and as you look up to see the source, there is a flash at the corner of your eye and you hear a squeal on the ground in the distance in front of you. Looking in that direction, the source of the shadow is revealed. Rising above the ground in a swirl of sand is a large, alien animal, sandy gray in color, with protruding eyes that look like glossy black mushrooms on stubby stems. The two articulated trunks rising out of its face are holding the hapless beetle-like creature you saw before. The hunter is a solid mass, with ruffles down both sides, a little like the kind you might see on the hips of a kindergartener as she dips her first toe of summer into the pool. The predator moves by undulating its sides like a cuttlefish, with a smooth and fascinating motion, carrying away its luckless prey. Death has come to this alien world.

The scene I've painted here has definitely never been seen by human eyes,

FIGURE 5.1. The plant and animal life of the Cambrian era is visually as alien as many a science fiction movie. While actual extraterrestrial life is likely to be much weirder than this, we can begin to understand the range of the possible by first looking at the great variety of life on Earth over the past half a billion years. © 2006 The Field Museum, Chicago. Illustrations by Phlesch Bubble Productions.

but it isn't fiction. Although my choice of the coloration of the plants and creatures came from our best scientific guesswork rather than knowledge, the scene I have painted for you comes from Earth's ancient history, below the shallow seas of the Cambrian period (figure 5.1). Plants as we think of them had not yet evolved, although single-celled algae had banded together into plant-like structures, and the sponges and corals of the era might have appeared like vegetation to the modern eye. The articulated beetle is a trilobite, of which there were many individual species, while the fourteen-legged worm bearing thorns is called *Hallucigenia* (figure 5.2). The fearsome predator of the early ocean with two trunks like a Siamese-twin elephant was *Anomalocaris* and could grow up to three feet long or perhaps more.

The biota of the Cambrian era is preserved in the Burgess Shale formation, located in the Canadian Rockies in British Columbia, as well as other locations around the world. It contains many creatures that are as weird as any Alien found in science fiction. *Opabinia* (figure 5.2)—with its articulated body, five eyes, tail like a modern fighter jet, and claw-like graspers that extend on a sinuous snakelike appendage as long as the rest of the body—would not at all look out of place in Hollywood's next blockbuster, set on a planet circling a distant sun.

This is *not* a book about the origins of life and the evolution that caused

FIGURE 5.2. The animal life of the Cambrian ocean includes body plans that have long since gone extinct, as shown in the arthropod-like *Hallucigenia* (*left*) and *Opabinia* (*right*). © 2006 The Field Museum, Chicago. Illustrations by Phlesch Bubble Productions.

the diversity we've seen over the past 500 million years. However, Earth is the only planet in the universe on which we are sure life exists. While alien life is likely to be totally different from Earth-based life, understanding the range of the forms of life that has existed on Earth is the first step in our exploration of what we might possibly encounter "out there." One thing I want to make clear is, as much as I dearly love the television show *Star Trek* and its spinoffs, it paints a totally improbable universe. Driven by the pragmatic need to have the characters played by human actors, the races in that universe are overwhelmingly humanoid. The chances are essentially zero that the Aliens we might one day meet in our exploration of the cosmos will be so familiar. Our visit to prehistory gives but the merest hint of how strange an alien world might be.

Lessons from Earth Life

Since before the days of Linnaeus, scientists have classified forms of life into different categories. Initially there were three classes, essentially animal, vegetable, and mineral (although the mineral class was quickly dropped). This early classification proved to be too limited to organize the staggering array of types of life that has been discovered. There are currently a couple of taxonomic systems in use which, for purposes of our discussion, aren't very different (no matter how passionately the various proponents debate them).

Biologists categorize living things according to their characteristics. The defining characteristics can be genetic or by form, with the two systems having considerable, although not complete, overlap. As an illustration, I offer a popular classification scheme. At the highest level are the domains, which distinguish life into Bacteria, Archaea, and Eukarya. The first two are grouped together as prokaryotes, meaning their cells do not have a nucleus. Eukarya just means that the cells of the organism do have a nucleus. It includes the kinds of life you see when you look out your window. Plants, animals, and fungi are each a different kingdom within the Eukarya domain. These kingdoms are further subdivided into phyla, classes, orders, families, genera, and species. Just to give you a sense of how each level distinguishes between the alternatives, we can see how our own human species is classified. Beginning with our location in the Animal kingdom, we then belong in the Chordata phylum (indicating that we have a hollow channel through which nerves can run—essentially the spinal cord). The next subdivision is the class, with our being a member of the Mammalia class. This means (among other things) that we are warm-blooded, hair-bearing vertebrates with females that produce milk. This classification is followed by our membership in the order Primates and then the Hominidae family and finally a genus and species of *Homo sapiens*. Since we are concerned with a broad sweep as we investigate body types, the details of these last few distinctions aren't so interesting to us.

Any self-respecting biologist would cringe over this cavalier description of an intricate system, hard won by centuries of effort. And they should. The painstaking interlinking of the world's species, finding which one fits in here and which one there, is a marvelous achievement. Indeed to have understood the tapestry of life and how species come into existence and live and die has to be one of the most successful achievements of science. When one includes the more recent genetic work, one cannot help but be awed by the story of life on this planet and even more by the fact that mankind has been able to figure much of it out.

However, we do not aspire to anything so grand here. We are interested in Aliens, not alien life per se. Recall that, in our context, "Alien" (denoted with the distinguishing capital letter) means a creature that can design, build, and fly a spaceship and with whom, in principle, humanity might someday vie for galactic domination. It is not critical that the technological levels be comparable, nor is it critical that they *actually build* a spaceship. Creatures who fly UFOs to Earth to invade are an inarguable example of what we mean by Aliens, but so are the Na'vi of James Cameron's *Avatar*. So Aliens must be mobile, intelligent, and able to manipulate the world around them. They must be able *in principle* to eventually pilot a spaceship. It is insufficient simply to be life that evolved on a different planet.

So we don't need to know about the equivalent of an alien chimpanzee, nor do we need to know if the alien planet has a creature like a squid. What we want to know is when we encounter an Alien life-form, what are the ranges of possible forms it can take? For that, we must consider much broader questions. What is the skeletal structure of the Alien? Does it have an endoskeleton like us, or an exoskeleton like a lobster? Is it hot blooded or cold blooded? Does it have distinct sexes and how many? It is the answer to queries like this that that we hope to explore by studying Earth-based life. After all, we know on Earth that there are many answers to these kinds of questions. While Aliens are likely to be very different in detail, we can learn a lot about what is possible simply by looking around us.

So we start our investigation by studying the domains and kingdoms. Of the three domains, we will put off discussion of the Archaea until the next chapter. Archaea life utilizes radically different metabolic choices and properly belongs in a discussion about life that exists under very different conditions than the ones with which we are most familiar.

The first of the domains we will discuss is Bacteria, which are typically unicellular and don't have a nucleus. Could Aliens be formed by evolved bacteria? (And by this I mean forming multicellular life using cells that have the structure of bacteria.) The answer is probably no. It's a matter of energy. Energy is formed in the cellular walls, and the bacteria structure has a much less complex cellular composition. This results in a much lower amount of energy available to the cell to do the sorts of things that would need to be done to form an intelligent Alien. Even though bacteria can come together to work cooperatively, this particular form of life just doesn't generate enough energy to be a viable building block of an Alien.

In fact, this is a good time to look at what is necessary to decide whether a particular body plan or biochemical approach is a credible candidate for making Aliens. The most basic consideration is energy. Evolution and environment can and do exert a powerful pressure to shape the direction that life follows, but variation like that can only occur if there is adequate available energy. If there is not enough energy to do something, it won't happen. It's a little like cars. There are Model Ts and jalopies and Ferraris. Car designers have come up with a vast array of different types of automobiles. However, one thing common to all of them is the need for an energy source. As we think about the life on Earth and how it might or might not have evolved under different conditions into an intelligent Alien, we should keep in mind the question of energy constraints and cars. There are many car designs and possible energy sources (e.g. gasoline, ethanol, wind, solar, nuclear, etc.), but a car needs some kind of energy to run. No energy means no motion.

So, if the energy source is simply too small to evolve the sorts of properties needed by an Alien—for instance, intelligence, mobility, and the ability to manipulate technology—then is it impossible for that particular Alien to exist.

Eukarya

Since the energy generation mechanisms of Earthly bacteria are simply too low to evolve into an Alien, we turn our attention to Eukarya. Eukaryotes are more complex kinds of cells than bacteria. These cells contain within them even smaller structures, themselves embedded in membranes. The central feature of Eukaryotes is the nucleus that gives them their name, which comes from the Greek *eu* (good) and *karyon* (kernel). The Eukaryotes contain other organelles that are the source of the energy in the cells. Mitochondria are the organelles that provide energy to animal cells, while chloroplasts provide energy to plants. There is a vast body of knowledge regarding the form and function of eukaryote life, and we will only touch on it in the lightest possible way, dipping into the details only when absolutely necessary. It is important to remember that the details of Eukarya aren't crucial. However, the boosted energy-making capabilities of Eukarya are.

Since we know that earthly Eukarya can generate adequate energy, it is valuable to explore this type of life a little deeper. Eukaryotes are broken into four kingdoms. They are the three that were mentioned before, animals, plants, and fungi, along with protists. We sort of intuitively understand the first three from our common experiences. Protist is sort of a catch-all category of organisms that don't fit into the first three. Protists tend to be unicellular creatures and, superficially, seem pretty similar to each other. Indeed it was only in the early 1980s when the diversity of protists began to be appreciated. The understanding of the evolutionary interconnections of the protists is an active area of research, but their unicellular nature makes them unsuitable for making Aliens. For multicellular life, we need to turn our attention to fungi, plants, and animals.

Fungi

Due to their superficial resemblance to plants, Fungi were originally classified simply as part of the Plant kingdom. Further study revealed considerable differences; for instance they do not photosynthesize and their cell walls can contain chitin rather than the cellulose of plants. Chitin is the material that forms the exoskeleton of many arthropods and insects. In fact, recent genetic work has revealed that fungi are more closely related to animals than to plants, although they are a distant cousin indeed. Unlike plants, fungi eat other things.

So with regard to our question, is it likely that fungi might have evolved into an Alien? The answer to this is no. Fungi get their energy from very energy-poor methods. There is simply not enough energy available for fungi to adapt the behaviors necessary to be an intelligent Alien.

Plants

The question isn't so clear a priori when one considers plants. The kind of Alien we're talking about will have to move in some way, and plants are generally immobile. However, the science fiction and fantasy literature abounds with examples of moving plants, from the "Feed me Seymour" plant in *Little Shop of Horrors*, to Tolkien's Ents, to the Whomping Willow of Harry Potter fame, the triffids from *The Day of the Triffids*, and the monster from *The Thing from Another World*. Is a mobile plant actually possible?

The kingdom of plants is exceedingly rich, filled with towering sequoias, annoying dandelions, prickly cacti, and languorously waving cattails. The range of body plans is quite diverse. Surely motion has evolved somewhere in the past? Does the phototropic ability of plants to move toward sunny environments count? Or the sudden snap of the Venus flytrap? Could these simple behaviors evolve toward more energetic mobility?

I think the answer to these questions is actually pretty clear and can be answered on energetic grounds. But before we discuss that, we need to talk a bit about the differences between plants and animals (which we know could evolve into an Alien). Both are eukaryotes, with a nucleus. Plants have a cell wall, typically made of cellulose, which gives plants their structure in the absence of skeletons. In contrast, animals have a cell membrane. Plants are autotrophs, which means they make their own energy, while animals are heterotrophs, which means they consume energy from plants and other animals and adapt it to their needs. The power source of animals is their mitochondria, which are tiny structures inside the cell, while the corresponding source for plants are similar objects called chloroplasts. Chloroplasts are structures inside plant cells in which photosynthesis occurs, converting light into metabolically useful energy. Chloroplasts contain the chlorophyll that gives plants their characteristic green color.

Do carnivorous plants tell us anything? If plants can eat animals or insects, surely we can believe that the more outlandish and fictional plants are at least possible? Actually, it might surprise you to know that the Venus flytrap and other similar plants do not get any energy from their prey. They get nutrients only, in contrast to other plants, which extract nutrients through their roots. In fact, nearly all carnivorous plants have evolved to live in extremely low-nutrient environments. If these plants are moved to a more nutrient-rich environment, they typically die. The calcium from ordinary tap water can kill a Venus flytrap . . . essentially because the plant grabs and stores the minerals it needs like a starving person might gorge on the roasting pig it finds at an abandoned luau.

But carnivorous plants are quite rare. Of the half a million or so plant species, only a few hundred are carnivorous. This is because the driving focus of all life is to acquire enough energy to reproduce. Since the parts of the plants involved in predation are poor energy collectors, the plant pays a price by turning leaves (solar energy collectors) to other uses. Essentially these plants evolve this way out of necessity. Just like cacti have unusual specializations to live in a place with very low water access; carnivorous plants have evolved their unique capabilities in order to exist in a "nutrient desert."

In order for plants to evolve to have animal-like properties, they would need to gain a nervous system, sensory ability, and mobility. This takes an awful lot of energy. Since plants only get their energy from sunlight, we can do some quick estimates of how much sunlight is necessary to power a human. It's not that an Alien must be human, but it gives a sense of the kinds of energy requirements that are necessary for a creature "sort of like us."

The resting energy usage of an adult human is about 60 watts, about that of an ordinary incandescent light bulb. That's just sitting there, doing nothing but having your heart beat, lungs fill and empty, and all those squishy organs in your midsection doing the sorts of things they do to get you through the day; getting up and moving around takes even more energy.

So how much sunlight does it take to power the average coach potato? The amount of sunlight hitting the Earth's surface at the equator is about 1000 watts for every square yard (assuming the energy receiver is always hitting the sun face on). So that would mean that our hypothetical, equatorial, plant-biology-based, human-like, coach potato Alien would need about a square foot, always facing the sun. Of course, the sun doesn't shine 24 hours a day. It's not like our heart stops at night, nor does sunlight always hit straight on. So we would need perhaps twice as much sunlight-grabbing area to store up the energy for the night, plus a little extra to account for inefficiencies in storing the energy for their midnight snack. In fact, accounting for night and the fact an Alien wouldn't always be facing the sun, the average amount of sunlight a creature could expect to see is 200 to 300 watts per square yard. Therefore, including the most basic considerations, we might think in terms of having maybe a few square feet to collect sunlight just to live and not move. In order to gain enough energy to move around, maybe we'd need a bit more. A square about 2 feet on each side is a reasonable amount of area, so this sounds promising. Maybe mobile plant-Aliens are possible?

But there's a problem. Chlorophyll doesn't absorb energy with 100% efficiency. Chlorophyll can, theoretically, collect about 10% of the sun's energy. However, plants typically achieve an efficiency of only about a third or half of that. Thus a hypothetical plant-Alien would need to have a surface area of about a square 10 to 12 feet on a side. But, of course, a solid animal that size would have a much higher volume and therefore weight (and correspondingly higher metabolic needs). If you sit and mull this over for a short while, you begin to appreciate why trees and bushes have the shape they do, with a compact trunk and then branches and twigs to simultaneously minimize the mass and maximize the sun-collecting potential.

We shouldn't forget the fact that plants also need to have a deep root system to get at the water and minerals below the ground's surface. Uprooting, moving, and rerooting would be an energetically prohibitive affair. Over the hundreds of millions of years of evolution, no plant based on Earth biology has evolved animal-like locomotive abilities (or at least we see no evidence for such a plant in the fossil record). This suggests that the ability to move is not consistent with the limitations of gathering energy from sunlight.

However, the numbers mentioned above give us some idea as to what kinds of factors might change this conclusion. For instance, chlorophyll, with its 3 to 5% efficiency in collecting sunlight, isn't up to the task under an Earth-like sun. If some other (and more efficient) chemical accomplished the task of collecting sunlight, that would change the calculation. Another factor that might make mobile and intelligent plant Aliens more credible would be to evolve in an environment in which there is simply more energy in sunlight to absorb. Of course, more sunlight comes with increased temperature, which means one starts to need to worry about boiling the water in the plant's tissues. Finally, there is another option, which would be plants that were sessile for a long time, gathering energy and storing it in (perhaps) sugars or lipids. The plants might spend a week, a month, or a whole growing season collecting energy that would be used either to let the plant move or to give mobility to offspring. (Visualize a tree that drops a walking orange or something.) This sounds outlandish, but is it qualitatively different from the sleep or hibernation of animals?

In summary, the chances of us encountering a plant-based Alien who evolved in an environment similar to Earth's is improbable due to physical limitations. A mobile Alien that absorbs the bulk of its energy from sunlight is not impossible, but it will require a different chemical to transform sunlight to metabolic energy and possibly a higher energy environment to supply the sunlight. Mobile plants with alternating mobile and sessile phases are also possible.

We should keep in mind that heterotrophs (creatures that consume other creatures) have an advantage in terms of being able to simply exploit the energy gathered by others. Like on our Earth, we can imagine that there will be plants that consume and transform sunlight or chemical energy (discussed in the next chapter) and creatures that take advantage of that ability and consume the plants. Remember that a blade of grass works hard to convert light into grass, but a sheep can consume many blades of grass, thereby benefitting from solar energy gathered over a large area. Effectively the grass has become an extension of the sheep's energy-gathering area, without the penalty of having to carry it around with them. Animals can simply consume a lot of the energy that the plants have produced. This might be an insurmountable advantage, even on a planet where plant mobility is energetically possible. After all, if the plants have more energy, this just supplies more energy to the things that eat them.

Animals

Following our discussion of the limitations of plant-based Aliens, we now turn our attention to animal-like life forms. Almost certainly any Aliens will be based on different biochemistry, with a different "genetic" encoding scheme. However, we know for certain that (1) Earth-based animal life could produce an Alien-equivalent and (2) that animal life on Earth has taken a vast variety of different forms. So we can take a look at the range of life observed on Earth to learn something of the possible.

The Animal kingdom consists of several phyla. The phylum including humans is Chordata, which, roughly speaking, means "has a backbone or spinal cord." There are other phyla that do not have a central nervous system. Some (like sponges) do not have differentiated cells.

When considering which of the phyla of the Animal kingdom might have evolved into an intelligent, tool-using, species, there seem to be a few crucial attributes. Differentiated tissue would seem to matter, as well as some ability to manipulate the environment. A central nervous system protected by a spine like we have doesn't seem to be crucial. For instance, the octopus, which has no bones at all and a partially dispersed nervous system, can exhibit remarkably intelligent behavior. They can be taught shapes and patterns. They can be trained to open jars with food in them. In 1999, scientists filmed octopi in the wild digging halves of coconut shells out of the seafloor. They then carried the shells with them and used them to form a protective shelter. This behavior was invented by the octopi and not trained into them by humans. This highly intelligent tool usage should totally destroy any vertebrate-centrism one might have.

Even insects can show evidence for types of intelligence. Honey bees exhibit considerable ability to communicate. Using a kind of dance, a lone forager bee can return to the hive and tell other bees where a food source is located. The other bees can then go directly to the food source. This could be considered an extremely complex instinctual behavior, but researchers have found that the ability of bees to communicate depends on their getting enough sleep. By depriving bees of sleep, their communication dance becomes less accurate. This suggests a type of intelligence that could in principle grow into something more akin to human intelligence, as it does not appear to be purely instinctual behavior.

The phylum Chordata is the most familiar to us, consisting of fish, birds, mammals, reptiles, and amphibians. These are the classes of animals that exhibit the behaviors most consistent with intelligence. So, for the rest of the chapter, we will explore the spectrum of body types, mobility types, object manipulation strategies, and other ways in which organisms interact with the environment. As we will see, there are an amazing number of options. However, during this discussion, we must guard against Chordata-centrism and keep in our minds the fact that nonvertebrate animals exhibit capabilities that perhaps could have led to intelligent life in an alternate history of Earth.

Alien Considerations

There are many properties one might consider when thinking about what an Alien might look like, things like body symmetry, number of limbs, and size. The next few pages discuss about twenty such considerations, using lessons taught us by earthly life.

Body Symmetry

The most familiar symmetry is called *bilateral symmetry*. This symmetry means that the left and right side are mirror images of each other. This particular body shape is favored by most of the higher animals. However, it is not the only possible choice. Spherical symmetry, where the body is like a ball, is possible in a water environment but difficult to imagine on dry land, where gravity would distort the body, unless it was hard. Another common symmetry is radial symmetry. This is the symmetry of jellyfishes, anemones, and starfish. Starfish have five or more arms, demonstrating a special form of radial symmetry, and many jellyfish have a four-way symmetry.

A final form of symmetry is no symmetry at all. This would be a life-form with some kind of lumpy structure, with protrusions and blobs here and there. An example of Earth-life with this body type is the sponge. Given the range of types of body symmetries seen here on Earth, it is hard to guess what symmetries an Alien might have.

Number of Limbs

There are a great number of choices here. Tetrapods, as their name suggests, have four limbs. This includes mammals, birds, and most lizards. Snakes have no limbs at all, although they evolved from a tetrapod ancestor. Insects have six limbs, while spiders and octopi have eight. *Hallucigenia* had fourteen. Centipedes have 20 to 300 legs, while millipedes have 36 to 400, with one rare species having 750 legs. Prehistoric *Opabinia* had but a single appendage.

There appears to be little Earth life can tell us about the number of appendages a life-form can have. However, our restriction that this be an Alien to compete with humanity for galactic domination makes it seem likely that it must have at least one appendage to manipulate the world around it. This is not a restriction caused by life, but a restriction caused by the need to invent and manipulate advanced technology.

Size

Our experiences on Earth can't tell us much about the size we can expect an Alien to be. The size of animals ranges from tiny insects to giant whales. Other restrictions suggest that intelligent Aliens are unlikely to be wholly water dwellers, although an amphibious lifestyle or even semiaquatic species, such as seals and penguins, are possible. While whales and dolphins are intelligent, we must recall our definition of Aliens. Underwater species cannot exploit fire, which is necessary for a species to attain the technology level to qualify as an Alien.

The requirement of mobility on land makes very large animals unlikely. So whale-sized Aliens are improbable. We do know of rather large dinosaurs. This might set a reasonable upper limit on the size of Aliens.

At the smaller side, the issue is neurology and intelligence. Too small a creature and there is no possibility for individual intelligence to develop. The situation is confused somewhat by the concept of a hive mind. Individual bees or ants seem to have minimal intelligence, yet the collective behavior is actually quite complex.

Individual creature intelligence is observed in octopi, small primates, raccoons, and animals of similar size. This sets a rough limit on the likely minimum size of an intelligent Alien using Earth-based neurology; in the range of a small cat. With a different brain structure, this restriction might be removed.

Obviously any discussion of size is dependent on the gravity of the planet on which the Aliens formed and the type of skeletal structure that supports the equivalent of muscle tissue. A planet with a lower gravitational constant will allow larger creatures.

Skeleton

Any land animal will likely need a skeleton of some kind. The boneless octopus would have considerable difficulty with locomotion on land compared with an animal with some sort of structure. Common animal skeletons are endoskeletons (inside the body like birds, mammals, and lizards) or exoskeletons (surrounding the body like insects and lobsters). I can think of little reason for one versus the other, except that a creature with an exoskeleton will have to molt to grow. However, there are other options, including a mixture of both technologies, or a young form of the life that has bones which dissolve after maturity, when an exoskeleton is formed. While not having an exoskeleton per se, the turtle combines a hard outer shell with a traditional skeleton. And, of course, a skeleton needn't mean bone. Cartilage, chitin, and other substances could be employed.

Nervous System

According to legend, if you're ever attacked by zombies, you always go for a head shot. It's the only way to be sure. The reason for this is the central nervous system observed in mammals. We have a brain that is connected to the rest of the body first through the spinal column and then a branching network of nerves. This particular design has some convenient features, as it central-izes thinking and the motor control that governs the body. However, there is no a priori reason why a creature couldn't have a distributed nervous system, with bits of their equivalent of a brain spread out over the body. If we ever encounter such an Alien, we better hope that they don't become zombies.

Locomotion

There are a tremendous number of locomotion strategies employed by Earth life. There is walking, flying, swimming, slithering, hopping, tunneling, and brachiating. There are also animals that move on the surface of the water.

For swimming, there is the motion of a fish (with a tail side to side) or a dolphin (tail up and down). There is the use of flippers like a turtle and the propulsion of squids and cuttlefish. Swimming capabilities have indepen-dently evolved several times, resulting in similar, streamlined body shapes im-posed by the need to move quickly through the water.

Flying has evolved on Earth at least four times, with birds, pterosaurs, bats, and insects, suggesting that this is a fairly common locomotive adaption. A flying Alien is entirely plausible.

There is little reason to select any particular form of locomotion for our Aliens.

Speed

The speed of an animal is tied to many other things. For instance, a heavily armored animal is typically slower than one without armor. Predators tend to be fast. On the other hand, humans aren't particularly fast in the Animal kingdom. There is little the animal world can tell us about Alien speed.

Color

The color of animals spans the rainbow. Aliens could have any color.

Defenses and Offenses

The natural defenses and offenses that Aliens could have are quite broad. Hu-mans are actually rather unimpressive in their offensive and defensive skills, but make up for it in their ability to utilize weaponry to overcome their struc-tural limitations. Any Alien capable of building a spaceship will have similar skill. However, there is no reason that the Aliens will not have other abilities. In nature, animals exploit a myriad of defensive and offensive strategies, from the camouflage of a leafy sea dragon, a tiger, or a cuttlefish to the venom of a cobra, a scorpion, or a male platypus. Mammals tend to not be venomous, perhaps because they are quick enough to kill with tooth or claw, whereas venom takes time.

Shells, horns, and spines provide protection, for example the tortoise and the ankylosaur or the porcupine and the blowfish. And, of course, simply avoiding the conflict through a burst of speed is a wise defensive choice. Rabbits, the spine tailed swift, and the gazelle are able to move extremely rapidly.

Temperature Regulation

The internal temperature of an animal can be regulated by the body's own metabolism (endothermic) or can depend on the environment, as is the case for insects, fish, and reptiles (ectothermic). In general, internal temperature regulation is a safer evolutionary choice, given that ectothermic species can be sluggish when the environmental conditions are colder. However, there is no reason to expect that the Aliens come from a planet as cold as Earth. Perhaps their planet is sufficiently warm that there is no need to evolve endothermy. Since metabolism depends on enzymes that tend to operate best in a fairly narrow temperature range, endothermic animals generally have a considerable advantage, but, in the right environment, the selection pressure may be small.

Blood

Blood is not necessary for all animals. Some insects use a fluid called hemolymph to transport oxygen to their tissues. However the higher animals use a substance that enhances the oxygen-carrying capacity of the liquid inside them. The most familiar type of blood contains a compound called "hemoglobin," which contains iron and gives blood its red color. Each hemoglobin molecule can attach to up to four oxygen molecules and increases blood's oxygen-carrying capacity to more than seventy times what it would be if the oxygen were just dissolved in water.

However, the iron-based hemoglobin molecule isn't unique. There are other options. For instance, some insects have a copper-based blood, using a compound called hemocyanin. Hemocyanin transports oxygen about a quarter as efficiently as hemoglobin, so it is more suitable for creatures with lower metabolic requirements. Oxygenated blood containing hemocyanin is blue. And sea squirts and sea cucumbers carry a vanadium-based protein in their blood called hemovanadin. There is still some controversy on the role of this protein in oxygen transportation. When oxygenated, it turns mustard yellow. It is pale green otherwise.

Diet

It is, of course, very difficult to know the diet of an Alien, but here on Earth, there are three options: carnivore, herbivore, and omnivore. Carnivores eat meat, herbivores eat plants, and omnivores eat both. There are pros and cons to all three. Herbivores have the greatest access to food, as plant life is ubiquitous. However, plant food tends to be lower in terms of calories available, leading to frequent eating. Carnivores have fewer food choices, as they must catch and eat other animals. This places constraints on their bodies, from the trapping of spiders, the "lurk and grab" of alligators, the "stalk and pounce" of cats, and the "communal attack" strategies of wolves. Carnivores get substantial nutrition influx with a successful attack, but they don't have as reliable a food source as herbivores.

Omnivores (of which humans are a member) get the benefits of both food sources. It is hard to imagine an intelligent Alien that doesn't have at least omnivorous capabilities, although it may opt to utilize one food source more often. We should also keep in mind that it is possible that the Aliens might need certain minerals or other substances, similar to Earth-life's need for water and salt. It is therefore possible that the Aliens might need to ingest materials directly from the ground, like deer around a salt lick. Given that the Aliens will evolve in an ecosystem with a common biological heritage, it is likely that some of this mineral collection will be done by plants for subsequent ingestion by Aliens.

Respiration

Respiration is the intake of vital gases from the environment (oxygen in the case of most animal life on Earth) and the removal of waste gases (predominantly carbon dioxide). As we will discuss in the next chapter, Aliens might elect to use different molecules in their metabolic processes, but the mechanisms to exchange gases with the environment are likely to be similar, as the phenomenon must satisfy basic physical constraints. These constraints include collecting gases from the outside and dispersing them to the tissues of the body. The respiration system is likely to be internal; otherwise something could block the ability to breathe. (For instance, imagine if your lungs were on your outside and you somehow got splattered with mud.)

Small insects have the simplest respiratory system, by exploiting the diffusion of gases into and out of the circulatory system. Recent research has shown that insects have a diverse range of respiration techniques, with some using muscles to expand and contract their respiratory systems in ways that are not terribly different from higher animals.

Land animals usually use a lung system, with an intricate system of branching pathways. The inside of a higher animal's lungs looks a little like a tree and for essentially the same reason. This design maximizes the area to exchange gases in the minimum volume. While birds, reptiles, and mammals differ in detail, the basic structure is similar.

Water-breathing animals, like fish and mollusks, use a gill system to extract oxygen from water. Extracting oxygen from water is a tricky business. Water contains about 3% the oxygen held by an equivalent volume of air. Consequently, fish have evolved highly efficient gills to extract approximately 80% of the oxygen from the water. (This can be contrasted with an approximate 25% extraction efficiency for mammals breathing air.) Still, this scarcity of oxygen might make it more difficult to evolve highly intelligent Aliens under water. Amphibians have a split system, breathing both through lungs and their skin. This ability to breathe through their skin is of great value when submerged in oxygenated water.

Environment

Does the Alien live on the ground, under the earth, under water, or in the air? This is one of the questions for which we can likely exclude some options. While animals exist in all of these environments, it is essentially impossible for our Alien to be purely a water breather. The reason is that we impose the need to have an ability to build a spaceship. While it is clear that intelligence can exist underwater (e.g., dolphins and octopi), building a spaceship requires technology, specifically manipulation of metal. It is very difficult to imagine an advanced technology that doesn't shape metal. Forming metal requires heat, which means fire. Since fire is impossible under water, it seems that our Aliens cannot be (solely) water breathers. An underwater alien caveman is possible. An Alien in the sense we mean in this book is not.

Reproduction

The number of reproductive strategies employed by animals is astonishing. There is the sexual reproduction of higher animals and the asexual reproduction often seen in microscopic organisms. Some creatures can do both, that is, reproduce sexually or asexually, depending on the environment. Asexual reproduction creates clones of the parent, which have the same susceptibility to disease or environmental change. Sexual reproduction ensures the genetic material is mixed. This results in a more diverse gene pool and is a guard against a change in the environment that might kill one individual but for which others might be better adapted. And, of course, for sexual reproduction there is external and internal fertilization, as well as egg-laying versus live birth.

Some species produce many offspring, knowing that many will not survive to reproduce themselves. An example might be frogs or rabbits. Other species produce fewer offspring but spend more time with them to ensure that they survive. This is the evolutionary tactic taken by humans.

For some species that have sexual reproduction, there are hermaphrodites, whereby a creature has the reproductive organs of both sexes and can both impregnate others and bear the young of their species. There are also species with tremendous sexual dimorphism, like the angler fish, in which the male fuses itself to the female and then atrophies away until he is nothing more than a sperm source.

An unusual adaptation in a few species actually has more sexes than the usual two. There are species in which individuals change from male to female and back again. There are species in which there are large "alpha" males with harems and smaller males of the same species with coloration that mimics the females. They hide in the harems and reproduce that way. There are insects in which a single dominant female lays the eggs and the other females are reproductively neuter. Even on Earth, sex can be complicated for a species. There is no reason to believe the male/female dichotomy will apply to Aliens.

Senses

What senses will our Alien have? It seems that a sense of touch is crucial to essentially all living organisms. Having a tactile awareness of your environment is important, whether you are predator or prey if, for no other reason than to know if something is biting you. Hearing is similar to touch, although there is a broad variation in how well species can hear. Taste or something similar allows organisms to decide if something is food or not. Vision is a very important sense and has evolved independently several times. Vertebrates, cephalopods (e.g., squids), and cnidarian (e.g., box jellies) have "camera-like" eyes, and each followed a separate developmental history.

There are at least ten different "eye technologies" that probably originated from a small spot of photoreceptive proteins on a unicellular common ancestor. However the details vary, from the human-type eye, in which focus is accomplished by changing the shape of the lens, to another choice in which the lens doesn't change, but the shape of the eye does. Then there are the multiple lenses of insects, the reflective eyes of scallops, and many other designs. Thus, while the details of the vision might be quite different, we can conclude that it is probable that our Alien will be able to see. It is simply too valuable an adaptation in a lighted environment to do without.

Of course, by "see" we don't mean just "see what we can see." Some snakes are able to detect infrared. Birds, reptiles, and bees can see some ultraviolet. So the possibilities of Alien vision are quite diverse.

It is important to recall that much of the vision of Earth creatures is optimized to see light where the sun is brightest. Aliens evolving on another planet would likely evolve the ability to see best using the brightest light available on their world. Thus it is possible that they could see the kind of light we do only poorly.

Senses that some Earth-life has that humans don't include the echolocation of bats and dolphins (useful in low-light environments), the ability to sense electric fields like some fish and sharks, and the magnetic sense of many migratory species (e.g., some birds, tuna, salmon, sea turtles, and more). We can also imagine Aliens developing a sensitivity to radio waves.

Obviously it is not mandatory that Aliens will have all of the senses we do. For instance, a subterranean species would have no need to develop sight. Tactile and auditory senses seem like they would be universal, as they would be helpful in any environment. A sense of smell or taste provides a method of chemical analysis; for example, some poisons taste or smell bad. Both senses might not be crucial, but having one or something similar would probably provide an important survival advantage.

Communication

The communication between Aliens will be aligned to their senses. Here are some options that Aliens might exploit: motion, smell, light, sound, or radio. Imagine trying to talk to an Alien who uses scent to communicate. (Given how slowly smells travel and dissipate, this is an improbable scenario, but it helps one think about how difficult human-Alien communications might be.)

Life Span

This is difficult to generalize from Earth life. Mice live only a few years, while some tortoises might live to about 200 years. There appears to be no strong correlation with metabolism rates on Earth. But, given the many factors that go into determining longevity, it is difficult to predict an Alien life span, except to state that an Alien must live long enough to learn the technology of previous generations.

Social Structure

Animals spend their time in many ways, from packs, to herds, to a solitary lifestyle. It is likely that Aliens will be social creatures in a way at least somewhat analogous to humans. The need for communication and retention of technical knowledge over the generations almost guarantees that the individuals will work together.

Wrap Up

These attributes of life are certainly not intended to be encyclopedic but rather to give a flavor of the kinds of variation possible should alien life evolve using carbon as the basic building block and with a biochemistry that is similar to our own. Of course, on a different planet, with different sunlight and chemistry, life might be quite different. Exploring some of these other options is the goal of the next chapter.

In summary, the study of biology on Earth certainly teaches us something of what is possible when discussing what an Alien might be like. Surely this brief survey has not explored all the possibilities. It is also clearly very Earth-centric. However, it does show some of the range of what we might encounter. While we realize that our conversation here does not span all possibilities, we might close with the following thought: Knowing something is better than knowing nothing, as long as you know it's not everything.

SIX

ELEMENTS

The third planet is incapable of supporting life . . . Our
scientists have said there's far too much oxygen in their
atmosphere.

Ray Bradbury, *The Martian Chronicles*

In the previous chapter, we looked at what sorts of lessons familiar Earth life
might tell us about what an Alien might be like. These observations were
not meant to be exhaustive, as they were based on a very limited range of
biochemistry. Animals breathing oxygen and converting glucose into energy
and plants converting sunlight doesn't even span the range of observed biochemistry here on Earth, let alone the range of the possible. There are creatures on Earth that use methane to exist and others who extract energy purely
from chemicals, rather than exploiting (directly or indirectly) light from the
sun. Then there is sulfur respiration and fermentation, just to name a few
alternatives.

At the end of this chapter, we will talk about more "exotic" forms of Earth
life. Our real interest is about Aliens who could potentially visit our planet,
but their story is inextricably tied up in the question of non-Alien alien life.

One must have the second to have the first. Accordingly, we will spend some time exploring what we know about alien life and the limitations placed on such life by simple considerations of chemistry and physical law.

The reader should be aware that any writing on this subject is bound to be incomplete. As noted popular science essayist and pioneer geneticist J. B. S. Haldane wrote in his 1927 book *Possible Worlds and Other Papers*, "The Universe is not only queerer than we suppose, but queerer than we can suppose." It is quite reasonable to suppose that the universe will have a trick or two up its sleeves and we will be surprised more than once. Still, we can talk about what we know about the relevant chemistry. If nothing else, we will learn what the important considerations are for modern astrobiology.

What Is Life?

This question is seemingly so simple, and yet it has vexed some of the most knowledgeable scientists and philosophers for decades. While hardly the first writing on the subject, physicist Erwin Schrödinger's (of Schrödinger's cat fame) 1944 book *What Is Life?* is one such example. It is an interesting early attempt to use the ideas of modern physics to address the question. Both James Watson and Francis Crick, codiscoverers of DNA, credited this book as being an inspiration for their subsequent research.

The definition of life is not settled even today. Modern scientists have managed to list a series of critical features that seems to identify life. A living being should have most, if not all, of the following features:

- It must be able to regulate the internal environment of the organism.
- It must be able to metabolize or convert energy in order accomplish the tasks necessary for the organism's existence.
- It must grow by converting energy into body components.
- It must be able to adapt in response to changes in the environment.
- It must be able to respond to stimuli.
- It must be able to reproduce.

These features distinguish it from inanimate matter.

While these properties can help one identify life when one encounters it, they don't really give us a sense of the limitations imposed by the universe on what life might be like. The purpose of this section is to get a sense of whether a would-be science fiction writer is being ludicrous when he or she bases a story around an Alien with bones made of gold and liquid sodium for blood. So what does our current best understanding tell us that life requires? A combination of theory and experimentation suggests that there are four crucial requirements for life. They are (in decreasing order of certainty):

- A thermodynamic disequilibrium;
- An environment capable of maintaining covalent interatomic bonds over long periods of time;
- A liquid environment; and
- A structural system that can support Darwinian evolution.

The first is essentially mandatory. Energy doesn't drive change, rather energy differences are the source of change. "Thermodynamic disequilibrium" simply means that there are places of higher energy and lower energy. This difference sets up an energy flow, which organisms can exploit for their needs. It's not fundamentally different from how a hydroelectric power plant works: there is a place where the water is deep (high energy) and a place where the water is shallow (low energy). Just as the flow of water from one side of the dam to the other can turn a turbine to create electricity or a mill to grind grain, an organism will exploit an energy difference to make those changes it needs to survive.

The second requirement is essentially nothing more than saying that life is made of atoms, bound together into more complex molecules. These molecules must be bound together tightly enough to be stable. If the molecules are constantly falling apart, it is hard to imagine this resulting in a sustainable life-form. It is this requirement that sets some constraints on which atoms play an important role in the makeup of any life. Hopefully after this discussion, you'll understand the reason for the oft-repeated phrase in science fiction "carbon-based life-form."

Requirement number three is less crucial; however it's hard to imagine life evolving in an environment that isn't liquid. Atoms do not move easily in a solid environment and a gaseous environment involves much lower densities and can carry a far smaller amount of the atoms needed for building blocks and nutrition. Liquids can both dissolve substances and move them around easily.

Finally, the fourth requirement might not be necessary for alien life, but it is crucial for Aliens. Certainly multicellular life or the equivalent will not be the first form of life that develops. The first form of life that develops will be of a form analogous to Earth's single-celled organisms (actually, most likely simpler . . . after all, modern single cell organisms are already quite complex). In order to form species with increasing complexity, small changes in the organism will be necessary. Darwinian evolution is the process whereby a creature is created with differences from its parents. The first thing that is necessary is that the organism survives the change. After all, if the change kills it, it's the end of the road for that individual. Once there are changes that both allow the daughter organism to survive and possibly confer different properties, selection processes become important. Creatures who subsequently reproduce more effectively will gradually grow in population until they dominate their ecological niche.

So let's talk about these ideas in a little more detail.

Thermodynamic Disequilibrium

The most important consideration for any form of life is the need for thermodynamic disequilibrium. This mouthful of an idea is simultaneously intuitive and counterintuitive.

If you tell someone that energy is necessary for life, you likely won't get any argument. Plants absorb sunlight, people eat food; the need for energy is self-evident. Yet the reality is a little more subtle. Energy has a technical meaning in science. Energy can be found in a thrown ball, a coiled spring, and a stick of dynamite.

However what life needs is not energy but rather an energy difference. If the energy is the same everywhere, this is not useful. What's useful are energy differences. To illustrate this subtle difference, consider a water reservoir held back by a dam (figure 6.1).

On the water side, everything is equal. While the pressure changes with depth, the uniformity keeps the water from moving around. It tends to stay put. However the water has a kind of energy that scientists call "potential energy." (Potential energy is the kind of energy where something would move if we let it, like how the water would move if we broke the dam or how an arrow would fly from a stretched bow if the string were released.)

Now imagine that there is a hole in the bottom of the dam. Water would rush out from the water side to the air side. This is in fact how hydroelectric power stations work. The moving water turns a turbine, which generates electrical power.

The crucial takeaway point here is that an energy difference (and a subsequent flow from high energy to lower energy) is central to the creation of electrical power and that this is true in a more general sense. This is what is meant when we say "thermodynamic disequilibrium." Thermodynamic means energy and disequilibrium means "not equal" or different.

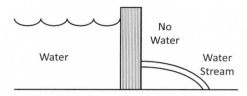

FIGURE 6.1. Water being held back by a dam is an example of an energy difference, and this energy difference can be converted into high-pressure water flow that can turn an electrical turbine. Although the energy differences of biology and biochemistry stem from concentrations of chemicals being held back by a cellular membrane, or in the interatomic bonds within molecules, the principle is the same.

Life works the same way. Energy differences allow energy to flow and make the kinds of changes that permit life to exist. For life, it is important to be able to store these energy differences for use when convenient. That way, an organism can move around, carrying its energy source with it. This provides protection against random occurrences that might restrict access to energy.

To get a sense of why this is important, consider a hypothetical alien cow that has to constantly eat to survive. If the cow exists on an ever-growing and ever-present patch of alien grass, there is no problem. However imagine a drought. With the death of the grass, the cow would immediately die, unable to move to a fresh patch of grass. Or imagine a plant that uses sunlight like Earth ones do but that can't store energy. It would live during the day, but die each night. Without a guaranteed, never-ending energy source, life of these forms is very vulnerable. Energy storage is necessary for life to exist.

It seems likely that life made of atoms (as we are) must exploit energy storage in molecules. Certain atoms can be combined together using available energy (as plants do with sunlight). Later, the energy can be extracted by converting molecules containing a lot of energy into lower energy ones and using the extra energy to live. We do this when we eat a cookie and metabolize sugars or fats. Perhaps an even more intuitive example of this would be when we burn wood. Cellulose combines with oxygen through a series of chemical reactions, resulting in carbon dioxide and water. We know that a fire releases heat—that's typically the point of fire after all—but what isn't so obvious is that what we're seeing when we toast our marshmallows is the transformation of molecules with lots of energy stored in their bonds into ones with less energy.

The Constraints Imposed by Atoms

Scientists know a lot about chemistry, how atoms interact and the properties of the matter they form. Surely this knowledge can tell us a lot about what elements are crucial for life. We are "carbon-based life-forms," as they say in science fiction. But science fiction also talks about other possibilities. The Horta in the *Star Trek* episode "The Devil in the Dark" was a life-form built around the silicon atom. Larry Nivens's Outsiders from his *Known Space* series have a biochemistry that includes liquid helium. Given the imagination of science fiction writers, both professional and amateur, I could imagine that sitting in somebody's drawer is a story about mankind's encounter with an intelligent race, with platinum bones and molten gold blood, who excreted diamonds. (If someone steals that idea and writes a story, I want a cut of the royalties.) So what does science tell us about the range of atomic combinations that is physically possible? For that, we need to think about some simple molecular requirements of life.

Life cannot exist without atoms combining together to make more complex molecules. Thus the way in which these atoms interconnect is a crucial consideration. While it may be obvious that the rules of chemistry are a defining aspect of any form of life, that statement is pretty vague. We can actually do better than that and discuss in the following text some detailed considerations.

For instance, alien life (and especially Alien life) will require a complex chemistry. Chemicals that perform analogous tasks to our familiar carbohydrates, proteins, DNA, and so on, will have to form molecules consisting of many, interlinking atoms. So two important considerations in the chemistry of life will be to identify atoms that (1) can make many connections to neighboring atoms and (2) can make strong enough connections so that the molecules are stable.

Students of chemistry have long been required to learn about valences, which is essentially the number of bonds the atom of any particular element can make. In order to make complex molecules, an atom will have to be able to connect to many nearby atoms. This can be made incredibly clear by considering the noble gas elements (helium, neon, argon, etc.), which inhabit the rightmost column in figure 6.2. These elements do not interact with other atoms. Each atom of the noble elements stands alone. They just don't participate in chemistry at all. Consequently, we can be certain that these elements

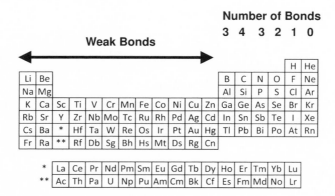

FIGURE 6.2. The atoms that make up matter each have a personality, with varying abilities to make stronger and weaker bonds and even different numbers of bonds. This variation between the elements is central to understanding all of matter, including life itself. Chemistry students will find the location of hydrogen (H) to be a little strange, being used to seeing it head the column that includes lithium (Li) and sodium (Na). However, each hydrogen atom can be seen as able to donate or accept an electron to form a bond, thus it could naturally be put in either location.

do not play a substantial role in any life-form's metabolism and certainly do not have a structural role in any form of life.

We can then consider the column immediately to the left of the noble elements. This column—which includes hydrogen, fluorine, and chlorine—consists of atoms that can form one bond with a neighboring atom. Since all of these elements act similarly, we can illustrate the point considering just hydrogen. It's kind of like a room full of one-armed people. They can hold hands with only one other person at a time. In a world in which hydrogen is a building block of life, you can only make very simple molecules, specifically ones consisting of identically two atoms. If hydrogen can form only one bond, then one atom of hydrogen bonds to a second atom. Both atoms form a single bond and the result is a two-atom molecule, as shown in figure 6.3. This is true for all elements in that column.

Moving one column to the left, we encounter the two-bond elements. The lightest example of these atoms is oxygen. Since oxygen can form two bonds, it can take on two hydrogen atoms. This is how water is formed, with an oxygen and two hydrogen atoms. Invoking our example of arms, oxygen is a two-armed element. It can hold hands with two hydrogen atoms or hold

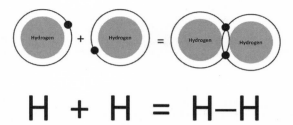

H + H = H–H

FIGURE 6.3. This is a couple of ways to represent how hydrogen atoms (H) combine to make a hydrogen molecule (H_2). The electrons of the two atoms are shared between them. On the bottom, we see a shorthand, with the atomic symbol standing in for the atom and a long dash (–) to represent the bond.

two hands with another oxygen atom. Moving again one column to the left, we encounter the three-bond elements. In a similar way, a nitrogen atom can connect with three hydrogen atoms and make ammonia.

However, the column that allows for the most intricate molecular structures is the carbon one. Carbon and other elements in that column can form four bonds. Continuing our exploration of bonding with hydrogen, a carbon atom bonded with four hydrogen atoms makes a methane molecule. In our analogy of arms, nitrogen has three arms, while carbon has four.

Carbon (like any atom) can connect with more than simply hydrogen atoms. It can combine with other carbon atoms, as well as all of the other atoms of the periodic table. Mind you, this is also true of the nitrogen and oxygen columns, but it is the ability to make four bonds that allows the most complex molecules to be created. Figure 6.4 gives just a sense of the kinds of structures that become available when one has atoms that have this many bonding possibilities. These are the molecules of life on Earth.

Now you've probably already gotten ahead of me and thought, "But what about the other elements in that column?" After all, silicon can also form four atomic bonds. Is silicon-based life possible?

Certainly silicon atoms can compose complex molecules; however the situation is more difficult than simply replacing carbon atoms with silicon ones. As a simple example, consider the common carbon dioxide that we exhale as we breathe. Carbon dioxide is a gas, which makes it easy for the fluid (i.e., blood) in our bodies to transport it. In contrast, silicon dioxide is a solid, known by the more common name of "sand." We will return to silicon-based life at the end of the chapter.

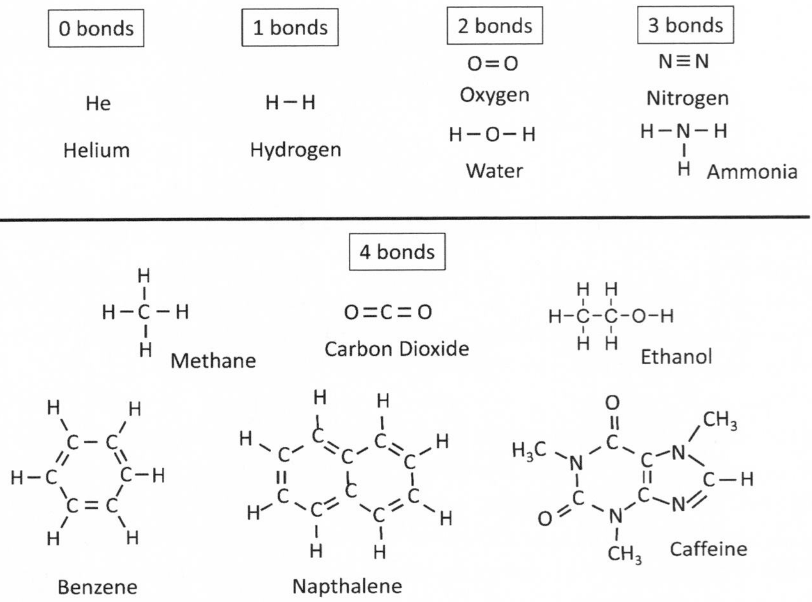

FIGURE 6.4. The different elements can participate in a different number of bonds, ranging from zero to four. The more bonds in which a specific element can participate has a large effect on the complexity of the molecules that can be formed.

Bond Strengths

While the number of bonds in which an atom can participate is a very important consideration, of equal importance is the strength of the bonds. The molecular and atomic world is a frenetic place, with constant motion being the norm. Due to simple heat, atoms vibrate, bounce into one another, and undergo a continuous stream of collisions. If the bonds aren't strong enough, these atomic and molecular collisions could rip apart the molecules of life, just like a hard tackle in football can cause a fumble. Without a stable molecular environment, surely no life could exist.

We can understand this point in a visual way by considering one of those reality television shows where they come up with ridiculous competitions. Suppose this show is called "Togetherness" and the point is that two people are tied together somehow and they are to stay together for the entire season. If their connection fails, they are disqualified. Suppose one couple is tied together with ordinary sewing thread, while another is connected with the kind of rope that mountain climbers use. It doesn't take much imagination to realize that the couple connected by a thread has a serious disadvantage. Just in the day-to-day to and fro of life, with walking around, brushing one's teeth, sleeping, and so on, something is going to break that thread. In contrast, there is very little that the rope couple will encounter that will cause them to be separated.

There are a couple of ways that atoms can be bonded together, but the strongest is called a "covalent bond." In a covalent bond, some of the electrons in each individual atom are shared between the two atoms. In a sense, the two atoms sort of fuse together into a single molecular unit. And these bonds are really strong. To give a sense of scale, two hydrogen atoms can bond this way to form a hydrogen molecule. The bond is so strong that if you took hydrogen gas at room temperature and pressure, you'd need a volume of gas the size of the Milky Way galaxy to have a 50% chance of breaking apart a single molecule into its two constituent atoms. These molecules are *really* hard to break apart. If they weren't, a volume containing that many atoms would have many broken molecules.

Getting back to the question of which atoms are most likely to have a significant role in life, we can ask if different elements can form stronger or weaker bonds. It turns out that the lower-mass elements can form much stronger bonds than the heavier ones. The reason is a little subtle, but luckily not too hard to understand. It all boils down to the degree to which the atoms overlap one another. The larger fraction of overlap, the more those two electrons are shared and the stronger the bond. This point is illustrated in figure 6.5.

This figure is simplified, but has some valuable features. Atoms consist of a nucleus and then a swarm of electrons around the outside. The electrons closest to the nucleus (or in the lowest energy states, if you've taken a chemistry class) are not generally available to form bonds, while the outer few electrons are. In figure 6.5, I've chosen to represent the core, noninteracting, portion of the atom as a black dot. The outer white circle is intended to represent the electrons available to form bonds. You'll note that I drew a small and large atom. For both atoms, the thickness of the white area is the same.

I then graphically made molecules, by connecting two atoms together. To a degree, one can say that the atoms share the electrons in the region between the two atoms where the white areas overlap. This overlap region is indicated in gray. Now compare the gray region to the white region in small-atom molecules and large-atom ones. You see that in the small-atom molecules that

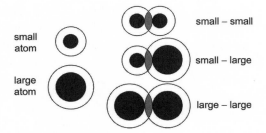

FIGURE 6.5. The strength of a covalent bond depends a lot on just how much the electrons from the atoms overlap. The larger the fraction of time they overlap, the stronger the bond. Here, the white area represents the electrons available for bonding, while the gray area represents the region of overlap. In smaller molecules, the gray area is a larger fraction of the white area.

the gray area is a larger fraction of the white area. Smaller atoms share their electrons with their neighbors a greater fraction of the time, which is the basis for the much stronger bonds in the lighter elements.

These simple considerations show why it is somehow natural for life to be formed of carbon. Carbon can form four strong bonds with neighboring atoms, allowing the formation of complex molecules. Other light atoms cannot form as many bonds, reducing the complexity of the possible chemistry, while other heavy atoms cannot form as strong a bond, thereby reducing the probability that the molecules will be stable. Carbon is an optimum element for complex molecular chemistry.

It is, perhaps, unsurprising that we carbon-based life-forms would conclude that carbon was an ideal basis for forming life. This is called "carbon chauvinism." We will return to this point when we've finished our overview of the important components of life and consider alternative chemistry.

Oxygen

All multicellular life on Earth uses oxygen as part of its respiration system, although this is not true of all forms of life. The role of oxygen is that it is a receptor of electrons. The movement of electrons is the source of the energy of life, so an element that can accept electrons is facilitating the flow of energy. Oxygen is a superlative acceptor of electrons.

Is the use of oxygen a necessary feature of life in the universe? Well, the answer is pretty clearly no, given that we know of life on Earth that uses other substances to breathe. In fact, we are quite confident that the first forms of

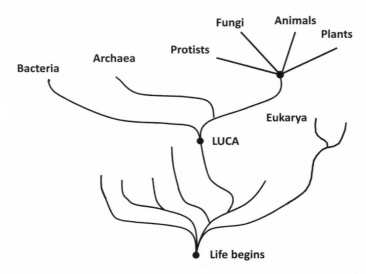

FIGURE 6.6. How it is believed that the first living organism began and underwent speciation is shown here. Eventually all of the early branches of life died out except for one organism that was the last universal common ancestor, or LUCA. This diagram shows only the most basic points, as cross-species genetic mixing is thought to have occurred when the organisms were simpler.

life on Earth would have been killed by the presence of oxygen. So what is it about oxygen and why has it become such a ubiquitous presence on Earth now? Does the universal usage of oxygen by multicellular Earth life mean that oxygen breathing is universal?

It doesn't, of course, but it's worth spending a little time learning about the essentials of the role of oxygen in the history of life on Earth. We don't know very much about the first life on Earth. Life formed and many species evolved and became more complex. As is usual with evolution, some species thrived, while others became extinct. It is thought that one of these complex organisms is the parent of all existing species, while the others died out. This parent being is called the last universal common ancestor, or LUCA. A family tree showing how life might have branched out is depicted in figure 6.6.

Working backward from today, biologists are quite confident that mankind shared a common ancestor with chimpanzees. That common ancestor shared an even earlier ancestor with other primates. The primates shared a common ancestor with other mammals. Moving backward in time, we now believe that each of the domains, kingdoms, phyla, classes, and so on, mentioned in the previous chapter originated from a common ancestor, whose descendants varied slightly and consequently set into motion the physical and biological differences we observe now in these different divisions of life. Each of the domains of Prokarya, Eukarya, and Archaea had a different common ancestor, although modern research suggests that Eukarya was formed by a mixture of earlier Archaea and Prokarya ancestors.

Taking the pattern one step further, there presumably was an organism who was the ancestor of all forms of life on Earth. Now this ancestor (the last universal common ancestor, or LUCA, mentioned above) was not the first form of life the Earth saw. Using comparative genetics and biochemistry, scientists have learned a lot about LUCA. For instance LUCA used DNA and a couple of hundred proteins to live. LUCA was already a very complex organism, quite different from the earliest form of life. It's hard to know which adaptation from LUCA gave it the edge to survive and thrive, while all of its cousin contemporaries were doomed to extinction. But survive it did and here we are.

LUCA probably didn't depend on oxygen for its respiration. While our understanding of LUCA's biochemistry is incomplete, it seems to be true that iron was an important part of its metabolic pathways. This fact is pretty conclusive evidence that LUCA lived before the Earth's atmosphere had a lot of oxygen in it. We know this as iron really loves to combine with oxygen into a form that is *extremely* insoluble in water. If there was a bunch of oxygen around, the iron would get gobbled up and pulled out of the ecosystem in the form of rust. As you've no doubt experienced, rust doesn't dissolve and, once the iron is in the form of rust, it is unavailable for future use. In order for an organism to depend a lot on iron means that it must exist in an anoxic (low/no-oxygen) environment.

While the date of the formation of life on Earth is an ongoing topic of debate, the period of about 3.5 billion years ago is a credible position, and the evidence grows increasingly stronger after about 2.7 billion years ago. Studies of the isotopic composition of early rock suggest that before about 2.4 billion years ago, there was very little oxygen in the atmosphere. However, at 2.4 billion years ago, the amount of oxygen in the atmosphere began to rise. The source of the oxygen was presumably early photosynthetic bacteria. For about half a billion years, the iron in the ocean absorbed oxygen and settled out on the ocean floor. This process went on until the iron was entirely used up and is the source of the iron mines we now exploit.

Once the iron was used up, the oxygen in the atmosphere began to rise much more rapidly. As I mentioned, the source of oxygen was photosynthetic bacteria that had existed since the earliest forms of life, but, given oxygen's reactive side, the oxygen was quickly bound to other substances in the ocean and eventually on the land. However, once these oxygen-loving materials in the sea and on the land were saturated, the oxygen concentration in the atmosphere increased. As the concentration of oxygen in the atmosphere grew, it encountered ultraviolet light from the sun. This led to the formation of ozone, which shields the Earth's surface from ultraviolet light (and makes land-based life possible). Without ozone's protection, the ultraviolet light would sterilize the surface of the planet, just like we use ultraviolet light to sterilize surgical instruments and to kill algae and parasites in fish tanks.

About 800 million years ago, the amount of oxygen in the atmosphere began to rise rather rapidly. This increase in oxygen is an oft-cited contributor to the origins of multicellular life (and, especially relevant to the idea of Aliens, animal life). The oxygen provided a large reservoir of a substance in the atmosphere that was an excellent acceptor of electrons and whose use in respiration and metabolism could generate lots of energy.

So oxygen is ubiquitous on Earth and plays a central role as part of all animals' energy budget. The question when we think about Aliens is "is oxygen *necessary*?" We know of life on Earth that uses other substances as electron acceptors, with ferric iron, nitrates, sulfates, and carbon dioxide to name a few. However, these alternative forms of respiration are found in microbes, not multicellular animals, suggesting that the benefits of oxygen respiration are substantial and that evolution will likely nudge biochemistry in that direction if possible.

Even on Earth, the mechanism whereby oxygen is used to give energy to organisms isn't a simple process but rather a multistep affair. Therefore it is possible that on a planet with an anoxic environment, evolution would invent a multistep process to get the required level of energy necessary to support Alien life. However, given the benefits of oxygen, it seems plausible that life would eventually find out a way to exploit it if it is present. This brings us to the next point.

Chemical Abundances

The chemistry we have been discussing is partially academic at this point. For instance, it may well be that carbon is the perfect atom from which to build life, but, if there is no carbon around, then it won't be used. Similarly, if there is no oxygen present, it makes it kind of hard to use it to breathe. So we need to add to our knowledge which elements are most present in the universe. To understand how certain elements are more or less common, we need to understand their origins.

Current theory is that the universe began just shy of 14 billion years ago in a cataclysmic event called the big bang. While the physics of the big bang is a fascinating topic, for our purposes, we merely need to know that the universe was once so hot that atoms couldn't exist; indeed individual protons and neutrons couldn't form, as the temperatures didn't allow them to coalesce out of the bath of energy and subatomic particles that existed at the time.

As the universe expanded, it cooled in a way that is analogous to more familiar explosions, and very early in the history of the universe, protons and neutrons came into existence, followed by the elements hydrogen and helium. For all intents and purposes, no other elements existed. Following our discussion above, life couldn't possibly form in that universe. Helium doesn't form molecules, and hydrogen makes simple molecules consisting of two atoms. If that were the whole story, we wouldn't be having this discussion. There must be more we need to consider.

Every morning as the sun rises, we are reminded of a seemingly trivial, but important, thing. The sun is bright and gives off heat. It does this because very dense collections of hydrogen and helium can undergo nuclear fusion. And nuclear fusion is one of the purest forms of scientific magic mankind has ever encountered and understood.

In medieval times, early scientists called alchemists were obsessed with the transformation of materials from one form to another; of "base metals" (e.g., lead) into gold. While there is no doubt that modern chemistry owes a debt to the early alchemists, they were doomed in their quest to transform one element into another. Such a goal is simply beyond the ability of chemical reactions.

However, the nuclear fusion of stars accomplishes just that. The nuclei of light elements are combined, forming heavier elements. In these stellar foundries, hydrogen and helium are forged into oxygen, carbon, nitrogen, silicon, and all elements lighter than iron. Standard, stellar-based, nuclear fusion fails to create heavier elements.

As it happens, some stars burn fast and furious and end their lives in a spectacular explosion called a supernova. In nearly a blink of an eye, these stars die, experiencing heat and nuclear reactions that dwarf those in more complacent stars. With their death, they form even heavier elements . . . even the creation of gold that eluded the ancient alchemists. This is the reason that

Carl Sagan so often stated that we are all "starstuff." Without stars, life and even planets would not be possible. In fact, the first stars formed when the universe couldn't have planets. The ingredients of planets just didn't exist. But, in their death, the early stars spread a complex brew of elements across the cosmos. These elements mixed with the existing hydrogen clouds and formed subsequent stars.

Our sun is a second- or third-generation star, having formed about 5 billion years ago. At the time of the sun's birth, the universe had had 9 billion years for earlier stars to manufacture the other elements of the periodic table. The elements present when our solar system came into existence formed the reservoir from which the planets and any possible life must be composed.

Figure 6.7 shows the relative abundances of the thirty lightest elements in our solar system. Hydrogen and helium make up 99.9% of the matter in the solar system, but from the remaining 0.1%, the planets coalesced. Of the remaining elements, carbon, oxygen, and nitrogen (the elements of organic chemistry and life as we know it) are the next most available. The relative abundance of all of the elements is in pretty good agreement with our understanding of how they are formed in the stellar furnaces in which they were created. Silicon, which is carbon's chemical cousin, is present in quantities that are about 10% that of carbon. So, a naïve interpretation of this graph might make you say, "well, yeah, it makes sense that life would be made of carbon, since there's more of it." Conversely, it doesn't take a lot of thinking to say, "Hey wait a minute. If carbon is so much more prevalent than silicon, why is Earth a big rock (i.e., silicon dioxide) rather than being made mostly of carbon? What gives?"

And, of course, that is an interesting question. The question of relative abundances of elements in the solar system tells us a lot, but life couldn't form from the elements inside the sun. It likely had to form on (or under or in the atmosphere of) the surface of a planet. So the correct elemental abundances to consider would be those on the surface of the planet. (The same logic that shows that the chemical composition of the star is only marginally relevant also rules out the molecular makeup of a planet's core as an important consideration. It is the makeup of the planetary crust that defines the reservoir of elements from which life can be formed.) I use the word "planetary" in a generic sort of sense. Life could have formed on moons of planets that are themselves sterile. We'll see in a short while the reason silicon doesn't play a central role in earthly life.

At this point, we begin to see how difficult it can be to generalize the dis-

Solar System Elemental Abundances

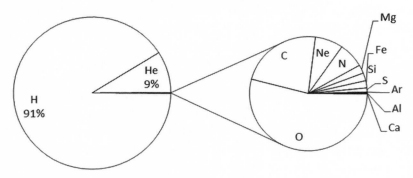

FIGURE 6.7. The distribution of elements in our solar system is totally dominated by hydrogen (H) and helium (He). Even the relatively common carbon (C) and oxygen (O) make up less than 0.1%.

cussion of chemistry and Alien life. After all, the environments on the various planets and moons in our own solar system are extremely diverse. The gas clouds of Jupiter are quite different from the scorched surface of Mercury, the frozen wastelands of Europa, and our own familiar Earth. It is this diverse range of environments that makes it so hard for astrobiologists to decide where to look for life.

However, we need to remember that we are interested in Aliens, rather than alien life per se. Aliens are creatures with sufficient intelligence to employ tools and someday compete with humans for galactic domination. Thus it is difficult to imagine a life-form suspended in the clouds of a gas giant as an Alien. It is much easier to imagine a creature on a rocky planetary object as a competitor. For one thing, access to metals is very important for manufacturing most tools and weapons. In a frigid environment, other materials might serve the same purpose. But in any case, the surface of a rocky planet is probably the relevant elemental reservoir to build our discussion of Alien life around.

We can start with the chemical makeup of the Earth's crust as a baseline. This is given in figure 6.8. There are striking differences in the Earth's elemental makeup compared with the solar elemental abundances, underscoring that the details of planet formation are critical. Hydrogen and helium are rare. We also see that the noble gases (helium, neon, argon, etc.) are noteworthy in their absence. These elements are gaseous and do not bind to other elements to form solids. Oxygen is the most present element, followed by

Earth Crust Elemental Abundances

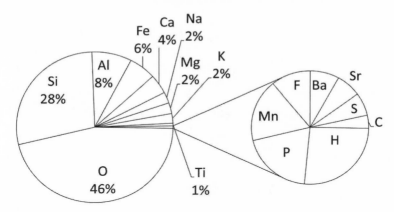

FIGURE 6.8. The elemental abundances of the Earth's crust reflect the fact that it is made of rock, which has a very high silicon (Si) and oxygen (O) component. The pronounced differences between the elemental makeup of the Earth's crust and the solar system as a whole highlight how accidents of planetary formation can significantly affect the chemical reservoir available to create life.

silicon. This mix reflects the various rocks (feldspar, quartz, etc.) that make up the surface of the Earth. Carbon is very rare in comparison to silicon (a tiny fraction of a tiny fraction, compared with about a quarter of the Earth's crust being made up by silicon). And this probably is telling us something significant. Even given the vastly larger amount of silicon available and the fact that both elements can create four bonds, life is formed from carbon. The ability to form four bonds is very important, but there are other considerations that must be taken into account when thinking about the chemical makeup of possible life. We will discuss at the end of the chapter silicon's issues as a building block of life. (I know I've promised this more than once, but we need a bit more background to explore the limitations of silicon as the basis of life, as well as to introduce an innovative way to overcome carbon's striking advantages.)

We will also talk a little later about the nature of liquid that forms life. On Earth, this liquid is universally water. As we wrap up our discussion of chemical availability, we can take a look at the elemental makeup of the Earth's oceans. This is given in figure 6.9. Because our oceans are made of water (H_2O), oxygen and hydrogen are the most prevalent atoms. Further, since most of the water on Earth is salty, it is unsurprising that sodium and

Oceanic Elemental Abundances

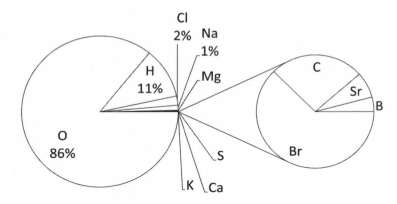

FIGURE 6.9. The elemental makeup of ocean water is a product of its chemical composition of water (H_2O) but also the fact that it contains salt (NaCl). Carbon (C) is a trace component of seawater.

chlorine, the elements that make up ordinary salt (NaCl) are present. The other elements are present if they can be tied up in molecules that are soluble in water.

As a final look at elemental availability, we turn to the human body. While the whole point of this discussion is to see what elements are available as building blocks of life, it is natural to ask "yeah, but what elements actually form life?" This is shown (for humans only) in figure 6.10.

Carbon, oxygen, hydrogen, and nitrogen dominate human chemistry, with a handful of other elements joining the mix. Our blood reflects our origins in the Earth's oceans. Calcium is used for bones and cellular metabolism. Trace minerals are found in our foods.

The fundamental question is whether other chemical compositions are possible for Aliens, and the answer must be yes. Biologists are still working out whether the composition of life on Earth is a historical accident or an inevitable consequence of the atomic properties of the elements and their relative abundances. Therefore, it is not at all surprising that astrobiologists haven't worked out what form Aliens or even the less restrictive alien life must take. But the limitations of chemistry and elemental availability are surely important considerations for their discussions. The topics we have discussed here—from the number of atomic bonds, to bond strengths, to elemental availability and evolutionary accidents and pressures—resulted in us. While

Human Elemental Abundances

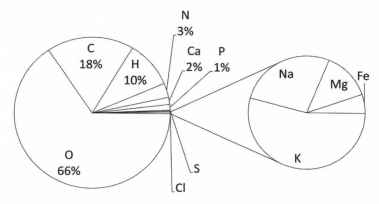

FIGURE 6.10. This plot shows the elemental abundances in the human body. We see why the crystal aliens in the *Star Trek: The Next Generation* episode "Home Soil" refer to humans as "ugly bags of mostly water." Given the chemical abundances of the Earth's crust and ocean, it is striking to see which elements are most present in living human tissue, with 97% coming from oxygen (O), carbon (C), hydrogen (H), and nitrogen (N).

oxygen breathing, carbon-based life-forms are not inevitable, we now see the advantages of that particular recipe.

Liquid Advantage

Life on Earth is universally water based, specifically liquid water. This naturally leads to two questions: Why liquid, and why water? The liquid question is the easier to answer. Matter typically exists in solid, liquid, and gaseous phases. The problem with the solid phase is the low mobility of chemicals. While solid-phase chemical mixing is possible, it is very slow. Life might form under those circumstances, but such life will never be an Alien in the way we mean it here. (Although we do need to keep in mind the idea of robotic life, as mentioned at the end of the chapter.) Further, unless the environment is totally dry, the advantages of liquid-based life are so manifest that either independently developed liquid-based life will out-compete the solid-based one or evolution will find a way for the solid-based life to adapt to using liquids.

In contrast, the gaseous phase of matter is supremely mobile. In fact, in many elementary school textbooks, a gas is defined as the phase of matter that fills up any volume into which it is introduced. So getting the gas molecules to move around isn't a problem. What's a problem is that a gas doesn't do a good job dissolving anything. While salt water can carry a goodly load of sodium and chlorine atoms, salt air only carries a little water, which itself contains the salt. So it is similarly unlikely that we will find life-forms (and especially Aliens) with a gaseous solvent.

So this leaves liquid. Liquid can move easily and is able to dissolve substances in it to move them around, like the salt in salt water. In order for a liquid to be a useful solvent, it must have two properties. First, to be useful a liquid must stay liquid under many conditions, and a clear implication is that the substance must exist in the liquid state over a large range of temperatures. Second, it must be able to dissolve and transport other elements. After all, inability to effectively transport other atoms was the reason that solid and gaseous solvents were rejected.

On Earth, the universal solvent of life is water. This miraculous substance may not be a universal solvent, but it is useful to discuss water's great properties so to understand what sorts of features other potential solvents must possess.

The covalent bonds we've already examined are not the only types of molecular bonds that are possible. Another important type of bond is called an ionic bond. While in a covalent bond, two adjacent atoms will share electrons; in an ionic bond, one atom will donate an electron to another atom. This causes one atom to have a positive charge and the other a negative one. The two atoms are then bound together by their respective charges. Common salt (sodium chloride) is like this.

Water molecules are an example of a polar molecule. This means that, even though they have no net electric charge, the electric charge inside them is not distributed equally. Thus one side of the molecule is, electrically speaking, "more negative," while the other side is "more positive." The interaction between the two sides of the water molecules and the molecules bound together with ionic bonds can break up the ionically bound molecules. In the case of salt, it isn't salt molecules that are present in water when you dissolve salt, but rather freely floating sodium and chlorine atoms. We see this in figure 6.11. This wouldn't be possible if water were not a polar molecule.

The electric charges from atoms set up electric fields, the means by which the atoms are attracted to one another. Water is able to shield electric fields very effectively, which is one of the reasons it can dissolve things so well. The dissolved atoms (say the positively charged sodium and negatively charged chlorine) are not able to see each other. Were they able to see one another, they would be attracted and recombine. This property of matter is called a

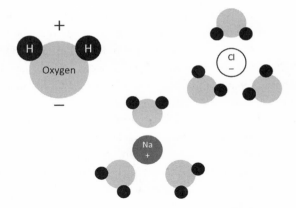

FIGURE 6.11. Water is a polar molecule, which means the arrangement of hydrogen and oxygen atoms causes one side of the molecule to have a slight positive charge, while the other has a negative charge. This property helps water dissolve materials held together by an ionic bond, like common salt, or sodium chloride (NaCl), shown here.

"dielectric constant," and it is very large for water with a numeric value of 80, which means that water can dissolve 80 times more of a solute than it would be able to otherwise. Water also can break up while in a liquid form, both donating and accepting a hydrogen atom, making OH^- (hydroxide, a base) or H_3O^+ (hydronium, an acid). The existence of acids and bases can be crucial for many chemical reactions relevant to life.

Water is liquid over a temperature range of 180°F, or 100°C. This range is quite large and will become important in a following chapter when we look at the concept of a planetary habitable zone. This is the range of distances from a star where the solvent (in our case water) will remain liquid.

Water has another hugely useful property. It takes a tremendous amount of heat to change its temperature. If you live anywhere near a coast, you know that the temperature at the beach is cooler in the summer and warmer in the winter than the surrounding areas. This is because on a terribly hot summer day, when the sun is beating down on you and you think you're going to melt, the water tends to be cooler than the air. While the sun is shining on you, it is also shining on the water. However water needs to absorb a (relatively) huge amount of energy to change its temperature, so it stays cool (and thereby cools the area near the beach, kind of like sitting next to the refrigerator with the door open). To assign a number, it is five times easier to heat up sand than water.

Similarly, in the winter, as a wintery north wind blows through you, biting cold, a large reservoir of water will contain considerable heat. This is the reason that the North Atlantic remains free of ice so far north, while the air is teeth-chatteringly cold. In a reverse of the concerns of the summer, because of the properties of water, the ocean has to lose a lot more energy to change its temperature.

Water has even more helpful and unusual properties. In addition to liquid water's being essentially a huge heat sponge, it takes a lot of energy to melt ice (and a corresponding large amount of energy must be given off to freeze water). Similarly, a large amount of energy is involved in converting water to steam, and vice versa. These properties are essential in the thermal regulation of the surface of the Earth.

Yet another curious feature of water is that, unlike most other substances, the solid phase of water (ice) has a lower density than the liquid phase. Basically, ice floats. Consider what would happen if the converse were true. When the weather got cold, ice would freeze and then sink to the bottom of the lake or ocean. As the ice descended, it would melt some but, in doing so, would cool the water below. Eventually the bottom water would be nearly the temperature of ice. Further melting and sinking would leave ice on the bottom of the body of water. After that, year after year, the ice would sink, building up the ice thickness until the lake or ocean was frozen solid, with only a small portion of the surface where there would be seasonal thawing of the water. The poles of the Earth would be frozen solid, from the ocean floor to near the surface.

However, real ice floats and insulates the water below from the colder air. Again, ice helps regulate the temperature of the surroundings. Without water, the Earth's environment would be very different.

Chemists have considered other possible solvents that at least have potential as a water replacement. One important consideration is the atmospheric pressure on the surface of the planet. We are necessarily somewhat biased, inasmuch as the pressure on the surface of the Earth seems normal. In contrast, the surface pressure on Venus is 92 times the pressure on Earth. At such pressures, other substances can be liquid over larger temperature ranges. For instance, on Venus, water can be liquid from 32 to 350°F.

For the following discussion, we limit ourselves to one earth atmosphere of pressure. At our familiar pressure, the following substances have been considered as possible solvents: water, ammonia, hydrogen fluoride, and methane (table 6.1).

TABLE 6.1. Comparison of possible solvents

Solvent	Liquid range °C (1 atm)	Density g/cm³	Heat capacity J/g K	Heat of vaporization kJ/mol	Dielectric constant	(Density solid)/ (Density liquid)
Water	0 to +100	1	4.2	41	80	0.9
Ammonia	−78 to −34	0.7	4.6	23	25	1.2
Hydrogen fluoride	−83 to +20	1	3.3	0.4	84	1.8
Methane	−182 to −161	0.4	2.9	8	2	1.1

Note: For the scientifically inclined, the units are: atm = atmosphere, mol = mole, J = joule, g = gram, K = Kelvin, cm = centimeter. Heat capacity is the energy needed to change the temperature of liquid, while heat of vaporization tells how hard the substance is to evaporate. The quoted densities are for the liquid form of the substance.

We can see the merits of the various materials. Ammonia has good thermal properties, but a limited temperature range over which it is liquid. On the one hand, hydrogen fluoride has a broad temperature range in which it is liquid, and it takes considerable energy to warm up the liquid, with the downside that it can convert to the gas phase very easily. It also has an attractively high dielectric constant. On the other hand, figures 6.7 to 6.10 show that fluorine is quite rare in the universe. Further, it reacts quickly with water to make hydrofluoric acid and with silicon-bearing rocks to make silicon fluoride. This is an inert material, which would tie up the fluorine and make it unavailable for respiration.

Note that methane is an interesting material; although not a polar solvent, it is a popular substance to consider when thinking about alternative biological chemistry. Methane can be found in its liquid form on the surface of Saturn's moon Titan for instance.

Hydrocarbons like methane have some advantages over water. Certainly empirical evidence suggests that the reactivity of organic molecules is comparably versatile in hydrocarbon solvents. However, since hydrocarbons are not polar, they are less reactive to some unstable organic molecules.

The surface of Titan is an excellent test case for many of these considerations. Titan is not in thermodynamic equilibrium, it has ample carbon-containing molecules, and it is covered with a liquid solvent. The temperature is low, which allows for a broad range of covalent and polar bonds. Indeed, it has many of the essential features that seem to be important to life. This leads us to speculate that if life is an inevitable outcome of chemistry, then Titan should have at least primitive life. If it turns out to not have life, then we must begin to suspect that there is something unique about the environment of Earth, perhaps including the use of water as a solvent. It is therefore not surprising that a probe to the methane oceans of Titan is a high-priority goal in NASA's exobiology plans.

Evolution Matters

The last property that seems to be necessary for alien life and definitely Aliens is some sort of Darwinian evolution. However life comes into being, it won't spring forth, fully formed, as an intelligent Alien, any more than it did here on Earth. Simple life-forms will be the beginning. They will encounter unstable environments, competition from members of the same species and others, predation, and so on. There must be a mechanism whereby organisms can change and adapt. If not, they will die out. It's just that simple.

However, precisely how this works is up for grabs. For instance, on Earth, the blueprint for life is stored in our DNA. Four nucleic acids—adenine, guanine, cytosine, and thymine—are the building blocks of the familiar spiral ladder of life. These nucleic acids make up the "rungs" of the ladder, while the sides of the ladder are called the backbone and consist of the sugar phosphoribose, which separates the rungs of the ladder.

Evolution occurs through a series of small changes that culminate in larger changes in the organism. The organism then competes in the ecosystem and may experience enhanced reproductive success. This is all pretty standard stuff.

What is a little more subtle is the realization that changes means just that . . . changes. It is imperative that the molecular structure that holds the genetic code be stable against small changes. The chemical properties of the DNA backbone must dominate the structure. Swapping a nucleic acid in or out must not cause the whole ladder to fall apart. This is critical. If the change causes the whole structure (and therefore organism) to be nonviable, then this is a disaster.

We can generalize these ideas beyond the specifics of DNA. The genetic molecules of any Alien must be able (1) to change without destruction of the molecule and (2) to replicate accurately with the new change. Self-replicating systems are well known in chemistry, but ones that can generate inexact copies, with that inexact copy also being faithfully replicable, are not. This might suggest that the genetic code of Aliens might need something analogous to the backbone of DNA, where the code can be "snapped in" like LEGOs. Surely the details of the molecules will be different, but the functionality is probably necessary.

Extremophiles

Extremophiles are organisms that live under conditions injurious to many forms of life. Now, from my observation, this should include people who enjoy being outside in Houston in August or a colleague of mine who summers in Antarctica, but extreme is actually quite a bit more extreme than that. Mankind has used extreme environments for a long time to preserve food. We now know that this is because these techniques kill or suppress the bacteria that would otherwise cause spoilage. A few techniques are to heat (i.e., cook) the food, refrigerate it, salt it, or even irradiate it.

And we all know this works. We have refrigerators and freezers. We have been admonished to cook rare roast beef to an internal temperature of about 140°F or as much as 180°F for well done beef or all poultry. The reason is to both cook the meat—to convert it from something raw to something yummy—and to kill the bacteria living in the raw meat.

There are other methods for preserving food that you have encountered in your local grocery store. There are dried vegetables, fruits, and meats, which have been starved of water, inhibiting bacterial growth. Nuts and other foods come vacuum packed to reduce the oxygen available in the package. Processing food by using high pressure can kill microbes. This is used for many products, including guacamole and orange juice.

Meat is cured by salting, as in the familiar bacon and ham. The high salinity kills germs. Smoking meats is also a way to store them. Sugar, even though it is rich in calories, is a good way to preserve fruits. Jellies and glacéed fruits can sit a long time without going bad.

Alcohol, aside from its mood-altering side effects, is also used to preserve some fruits. This is usually done in conjunction with using sugar as a preservative.

Changing the acidity or alkalinity of the food is another way to lengthen its lifetime. While salting plays a role in making pickles (and pickling in general), the use of vinegar (with its attendant acidity) can extend the shelf life of food. And, if you are of Scandinavian descent, you might enjoy lutefisk, which is fish prepared with lye, which is highly alkaline.

Atmosphere modification is also a useful technique. Food, such as grains, can be put in a container and the air replaced with high-purity nitrogen or carbon dioxide. This removes the oxygen and destroys insects, microbes, and other unwanted intruders.

The real point is that mankind has known about various ways to preserve food for millennia. Spoilage of food originates from undesirable creatures (typically microbes of some sort) "eating" the food and releasing waste products. Through some combination of the techniques mentioned above, we have learned to kill the undesirable bacteria that would otherwise ruin our food.

Our experience has led us to some understanding of the range of conditions under which Earth-like life can exist. However relatively recent scholarship has revealed that life is actually hardier than we thought.

Biologists have given the name "extremophile" (meaning "lover of extreme conditions") to organisms that thrive in environments that would kill familiar forms of life. While the study of extremophiles is still a fairly young science, we can discuss some of the range of conditions under which exotic life has been found.

At the bottom of the oceans, sometimes at extraordinary depths, there are spots where magma has worked its way from the interior of the Earth to the ocean floor. At these points, called hydrothermal vents, superheated water streams away from the magma. This water can be heated to well above the familiar boiling temperature of 212°F, but the huge pressure at the bottom of the ocean causes the water to stay in its liquid form. Water inside these hydrothermal vents can be nearly 700°F, certainly high enough to kill any form of ordinary life.

Only a few feet away from these vents, the temperature of ocean water can be very close to freezing, about 35°F. In this temperature gradient grows an unusual ecosystem. At the top of the food chain are relatively common types of clams and crabs who consume food in standard ways. However at the base of the food chain are thermophilic (heat-loving) bacteria that can live at temperatures above the usual 212°F boiling point of water. These bacteria do not use the same biochemical pathways of ordinary life. Rather than using oxygen as an electron receptor, they use sulfur or occasionally iron. These materials are spewed copiously into the sea, dissolved by the water from the magma source.

In fact, current thinking is that these prokaryotes are perhaps closest in nature to the last universal common ancestor (LUCA) of life on Earth. How could this be? Well, we should remember that LUCA was itself a sophisticated life-form and certainly not the only one around at the time it existed. While the following is purely speculation, we could imagine that this life-form might have survived a late strike on Earth by a comet or something similar. The impact would have vaporized the oceans and only the deepest-dwelling, most heat-resistant life might have survived.

Heat-resistant, sulfur-breathing life is not the only type that exists in extreme environments. On the other end of the spectrum are the cold-loving cryophiles. While pure water freezes at 32°F, salty water can remain liquid at temperatures much colder than that. Life-forms at the cold end of the spectrum have quite different problems compared with their thermophilic cousins. If water freezes, it expands and can rupture cell membranes. Plus the reduced temperature can significantly lower the rate of chemical reactions experienced by the life-form. In essence, cold life "lives slower." Further, just like cold butter is hard to cut, while warm butter is nearly a liquid, cold can stiffen the cellular membranes of cold life. Chemical adaptations are needed to mitigate the problems of the cold.

As of our current understanding, we know of no eukaryotic life that can exist at temperatures outside the range of 5 to 140°F. While the lower number is below the freezing point of ordinary water, water with high salinity can remain liquid at these temperatures. Microbial life has been observed over a temperature range of −22 to 250°F. An example of a cryophilic organism is *Chlamydomonas nivalis*, a form of algae that is responsible for the phenomenon of "watermelon snow," in which snow has the color and even the slight scent of watermelon.

Chemical considerations can give us insights into the ultimate constraints on the temperature of carbon-based life. Due to the bond strength involving carbon atoms, it's hard to imagine life at standard pressure much higher than 620°F; about as hot as the hottest your oven can bake. Of course, pressure can affect the rate at which molecules break apart and the decomposition of molecules can be slower at high pressure. It's probably safe to say that carbon-based life is not possible above about 1000°F at any pressure.

Water is critical to life, however it may be that there are extremophiles that don't need much of it. Looking for life in locations with little water is a way to better understand the realm of the possible. And Earth does have some extremely dry places. The Atacama Desert is commonly called the driest place on Earth. Some places in the desert get about a fraction of an inch of rain per year and some weather stations have never recorded any rain at all. There are tall mountains (over 22,000 feet tall), which one would expect to be glacier-covered, that are completely dry. In fact, there are empty river beds that have been estimated to have been dry for as many as 120,000 years. There are a few places in the Atacama Desert that are thought to be the naturally occurring place on Earth with conditions comparable to Mars. In fact, NASA has done some work there to help design Martian probes. They have gone so far as to experiment on searching for life in the sands of the Atacama Desert, using techniques that are hoped to definitively answer the question of life on Mars.

There are also forms of life that are halophiles (salt loving). In the Dead Sea region of the Middle East, most life couldn't survive. However, there are lichens and cellular life that have adapted their chemistry to maintain their inner environment in such a way as to thrive. Some of these forms of life actually need the high salt environment to live at all. It's hard to believe that an environment that can cure a ham is actually a comfortable place for life to live and yet it's true.

As with the other food-preserving extremes, life has been found in highly acidic and basic environments and even in the presence of radioactivity a thousand times higher than would kill the hardiest normal forms of life. These observations have certainly broadened scientists' expectations of the range of environments that life can successfully inhabit.

With the discovery of these extremophiles, scientists have intensified their search for the niches that life can occupy on Earth. We have pulled life out of well cores taken from a couple of miles under the surface of the Earth. Life has been found floating in the rarified air of the stratosphere. Microbes have been found as high as 10 miles above the ground. This environment is extremely harsh. The temperature and pressure is very low, the flux of ultraviolet light is very high, and there is nearly no water. Survival in this hostile environment inevitably raises questions of "panspermia," which is the premise that life might have arrived on Earth from some other body . . . perhaps Mars. While this seems improbable, it is not ruled out. But life had to start *somewhere*, so the questions we have discussed here are still relevant, even if life started elsewhere. Of interest to us here is the understanding that some primitive forms of life can exist in an environment that would kill creatures that live closer to the Earth's surface. However, this primitive form of life wouldn't be an Alien. But it does give us some additional information on precisely how resilient Earth-based life, with our carbon and water-based biochemistry, can be.

Silicon-Based Life?

In science fiction, there is soft SciFi and hard SciFi. In hard SciFi, the writer tries to advance the plot line constrained by the best-known science of the time, while in soft SciFi, more liberties are taken with the science. In the case of stories about alien life, a common alternative to our familiar type of life is one based on the silicon atom. The arguments presented earlier about the advantages of carbon (specifically the four bonds available and the rich chemical complexity that comes with it) are rather compelling, suggesting the four available bonds are a necessary condition of complex life. In fact, chemists have cataloged more molecules involving carbon than all the known molecules that exclude carbon. Think about that. If you took all elements except carbon and made every known compound, you'd have fewer compounds than the ones that have been found and contain carbon.

Given the benefits of four bonds, it is therefore natural that a hard SciFi writer who wants to break away from carbon-based life would then invoke silicon as the next candidate base element around which to build a fictional ecosystem. There's only one problem: it isn't as simple as that.

We've already noted the simple objection that while we breathe out carbon dioxide as a gaseous waste product, silicon dioxide is solid and we are more familiar with it as sand. This particular fact was noted early on in the 1934 short story *A Martian Odyssey*, by Stanley G. Weinbaum, in which he described a Martian silicon-based creature that excretes bricks every ten minutes. These bricks were the waste products of respiration.

However, the problems with silicon are much deeper and fundamental than this. Far more damaging are silicon's issues with its stability in its interactions with other atoms and the rate at which silicon chemically interacts.

A very important feature of how carbon bonds with other elements is that the bond strength between two carbon atoms (C–C) is quite similar to that of a carbon-hydrogen bond (C–H), as well as carbon-oxygen (C–O) and carbon-nitrogen (C–N). Because of this, it is energetically fairly easy for a reaction to swap out one atom and connect another. From an energy point of view, which of these elements participate in the bond doesn't matter much and so these swaps occur pretty freely.

In contrast, silicon doesn't have this property. It turns out that silicon-oxygen (Si–O) bonding is much stronger than with hydrogen (Si–H), nitrogen (Si–N), or even other silicon atoms (Si–Si). Consequently, silicon binds easily to oxygen (making silicon dioxide), and it is very hard to break apart that bond and slip in another atom.

What we've mentioned here is just a characteristic of single interatomic bonds. When we turn our attention to multiple bonds, carbon is again quite superior. It turns out that a double carbon bond takes about twice as much energy as a single bond, while a triple bond uses about three times as much energy. It didn't have to be that way. The details of multiple bonds are different from single bonds, and carbon just got lucky.

Silicon, in comparison, has a much harder time making double and triple

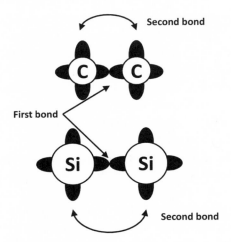

FIGURE 6.12. Because of their size and shape, silicon atoms have a hard time making stable double and triple bonds. The strength of the second silicon bond is much weaker than the first silicon bond. This is in contrast to carbon, in which the second bond is comparable to the strength of the first bond. The black areas represent electrons available for bonding. In silicon, the electrons participating in the second, third, and fourth bonds are separated by a greater distance and consequently bond more weakly.

bonds. This has to do with the size and shapes of the atoms. The pictures of figure 6.5 give an overly simplified impression of the shape of atoms. Silicon and carbon really look like spheres with bumps protruding out of them, with the bumps participating in the bonds. Because the silicon sphere is bigger than the carbon one, and the silicon bumps aren't much bigger than the carbon ones, the bumps are farther away between two adjacent silicon atoms. This makes it harder to get the bumps closer to other atoms to share electrons, which makes a second bond much weaker than the first one. Consequently, the strength of double bonds between adjacent silicon atoms aren't much different from single silicon bonds. This makes complex chemistry using silicon that much harder. This point is illustrated in figure 6.12.

Finally, the ease at which reactions can occur is much greater with silicon atoms. Consider a gas stove, inadvertently left on, so carbon-containing natural gas fills the house. The gas can fill the house, but it won't explode without a spark to set events in motion. However, a similar "silicon natural gas" would spontaneously react without the spark. This speed of reaction reduces the time necessary to form complex molecules.

So does this mean that silicon-based life is impossible? Could the rock people of planet X be having a discussion about the benefits of silicon-based life? Well, sure. It's not like the factors mentioned in this chapter are definitive, nor should you think that we've exhaustively explored all options. But these factors are certainly strong reasons to not think of silicon-based life as equally likely as other worlds covered with carbon-based life. Even Carl Sagan was reported to have stated that while he was only a weak water chauvinist, he was a huge carbon chauvinist.

So scientists must consider the possibility of alien life based on atoms other than carbon, but it isn't considered to be highly likely. However, when we talk in this way about silicon life, we need to remember that we've been talking about life that evolved directly from non-living substances. There is another form of silicon life that we should keep in mind.

Second-Generation Silicon

"Resistance is futile. You will be assimilated." This is one of the trademark phrases of one of the nemeses of humanity in *Star Trek: The Next Generation*. The Borg are cyborgs, which are a mixture of organic (i.e., squishy stuff like us) and cybernetic implants, which obviously include metals and silicon. In Fred Saberhagen's *Beserker* series, self-replicating robotic creatures roamed the cosmos, intent on destroying life. A computer called HAL in *2001: A Space Odyssey* became self-aware and turned on his crew. The eponymous Terminator is a self-aware robot tasked with the termination of humanity. The Cylons of *Battlestar Galactica* are at war with humans. The Daleks of *Dr. Who* wander around saying "exterminate." The silicon-based creatures of science fiction are often bad guys.

One can find many examples of cybernetic enemies of humanity in the science fiction literature. The story line is often similar to that of Frankenstein, when an artificial form of life gets out of control and turns on its creator. However organisms of this form must be considered life in the sense of how we mean Aliens. These cybernetic creatures (whether enemies or friends) would not have directly evolved from inanimate matter, but we should keep them in mind as we consider what sort of Aliens we might one day encounter. Indeed, when one considers a second-generation form of life—meaning one that is carefully designed by a first form of intelligent life (where by first form, I mean a type that has evolved from scratch)—many of the considerations listed here are less important. Metals, silicon, other elements could easily be essential parts of created life. Even second-generation carbon-based life could have a more complex and efficient biochemistry.

But, really, the idea of second-generation life is perhaps not the first concern for scientists looking for Aliens in the universe. However, if Alien spaceships ever appear over the cities of Earth, it's probably best to hope that they aren't in the form of big cubes. You know . . . just in case . . .

Wrap Up

While I've tried in this chapter to describe the most important considerations in the creation of life, you should by no means think that what I've said here is airtight. Some of the things are pretty inarguable, for instance, it seems exceedingly unlikely that helium will play a huge role in the biochemistry of Aliens. Helium just doesn't participate in atomic bonding. Further, there is a clear advantage to the use of carbon as a base element. Being able to create many bonds leads to a complex chemistry and a correspondingly diverse biology. It is also true that without adequate energy (and an exploitable energy difference), life cannot exist.

However, beyond that, it is hard to say anything definitive. Once one gets past the minimal chemical and physical considerations of life, evolution is a powerful optimizing tool. Earth-based biochemical cycles are extremely complex, and it is literally unbelievable that extraterrestrial biochemistry will not be both as complicated and different than the paths observed on Earth.

Still, we know enough about chemistry to know that some possible metabolic pathways cannot yield the same amount of energy as others. This does set some limits on the Aliens we might encounter. However, when we take into account that life might exist on planets with very different temperature or pressure than we find on Earth, the limitations are not quite as absolute as it may seem.

What I hope I've done is to have given you a sense that not all of the ideas you might encounter in science fiction are possible, for instance a sentient gas cloud is pretty hard to imagine. Still, the realm of the possible is still rather broad. Astrobiologists definitely have their work cut out for them.

SEVEN

NEIGHBORS

Sometimes I think the surest sign that intelligent life exists elsewhere in the universe is that none of it has tried to contact us.

Bill Watterson, *Calvin and Hobbes*

Thus far, we have discussed mankind's vision of Aliens without much attention to the contribution of modern observational astronomy. The first musings about the surface of Mars by Percival Lowell and his contemporaries were informed by the best science of the time, although we recall that many scientists dismissed his beliefs as ridiculous. However, you should recall that during that era, there were ongoing, multidecade arguments over the existence of canals on the face of Mars. These canals were reported to be thousands of miles long and to irrigate swaths of the planet's surface tens or even hundreds of miles wide. This tells us something about the technology available to those early scientists. By today's standards, it was crude. If cautious and sober scientists could imagine observing an extensive canal system on our nearest planetary neighbor, it was quite unrealistic for scientists of the era to have a more detailed and accurate picture of possible life on Mars.

This is not to say that the scientists of the early twentieth century didn't exploit all the instruments at their disposal. As we mentioned in chapter 1, those same scientists did have access to spectroscopy and so they were able to make crude measurements of the composition of some planetary atmospheres and determine the presence of substances vital to earthly life, like oxygen and water. For instance, most scientists of that era were aware that Mars was dry (hence the reason that the canal mythology resonated so well) and had very little oxygen. However, these early insights were but a first step along the path to modern planetology.

Continuing our review of what we have covered so far, in chapter 2 we looked at the impact of reports of extraterrestrials, mostly by observers who were not well-informed of the science of the day. Their stories, totally irrespective of whether they were objectively true reports or not, incorporated culturally familiar religious themes (Adamski) and nightmarish ones (Betty and Barney Hill). The science of the middle of the twentieth century mattered less to the public's vision of Aliens than did these UFO reports.

The science fiction discussed in chapters 3 and 4 is, by definition, speculative. Often science fiction writers have a respectable knowledge of contemporary scientific thought, but their goal is to tell a tale; often one that says more about humanity than it does about science or real Alien behavior. We should not trust these stories to be bound by the rigid strictures of the best scientific knowledge.

Even when we turned to more scientific thinking in chapters 5 and 6, this discussion was more about understanding the range of the possible, informed by what we have found here on Earth and later by the limitations imposed by the physical laws of the universe. While these are valuable lessons, what could be and what actually is are quite different things. Humans could have evolved to be 9 feet tall or could have descended from a nonprimate lineage. Neither turned out to be what happened.

Accordingly, in order to understand what real, true Aliens are, the only way we'll ever know the definitive answer is to either go and meet them and shake their hand or tentacle or whatever greeting is appropriate or talk to them somehow. Since we have no hard evidence that Aliens exist (the reports of chapter 2 notwithstanding), what do we actually know? What has science learned about the existence of extraterrestrial life (and, more importantly, Aliens)? What do we know about the probability of encountering life if we ever explore the galaxy?

Where Are They?

Enrico Fermi was a brilliant Italian physicist who is known to the public as the man who led the team that first harnessed nuclear power under Stagg Field in Chicago on December 2, 1942. His impact in physics was actually much broader than that, and he has been honored (among many other tributes) by posthumously lending his name to the Fermi National Accelerator Laboratory, America's preeminent laboratory for studying the basic building blocks of the universe. In addition to sheer brilliance, Fermi had a gift for trying to get at the bottom line, using simple estimators. We physicists call "a Fermi Problem" a question that is easy to ask, hard to know definitively, but able to be estimated by thinking it through. The most repeated example of a Fermi Problem is "How many piano tuners are there in Chicago?" By knowing the number of people in the city and then estimating how many households have a piano, how long a piano holds its tune, how long it takes to tune a piano, and the length of a work week, you can come up with a reasonable estimated answer. (Current estimate, about 125.)

Fermi lived in an elite academic world—an active mind surrounded by others of similar caliber. They would talk about all manner of things, looking at them from every angle, trying to get at the truth. From a casual lunchtime conversation, one of the most famous questions involving extraterrestrials was asked. The story goes something like this.

One summer day in 1950, Enrico Fermi was visiting the Los Alamos Laboratory, which had been the secret government facility at which much of the first nuclear weapons had been developed. He and three companions, one of whom was Edward Teller, were on their way to lunch. They were talking about a cartoon seen in the May 20 issue of *The New Yorker*, which explained a recent spate of thefts of trash cans in New York City as being perpetrated by Aliens taking them into their flying saucers. (The UFO mania of the late 1940s was still fresh in the public's mind.) The conversation then meandered to Teller and Fermi bantering back and forth over the chances of mankind exceeding the speed of light in the next decade, with Teller suggesting a chance in a million and Fermi guessing 10%. During the stroll, the numbers changed as they intellectually fenced.

After sitting down to lunch, the conversation went in a different direction, with Fermi sitting there quietly. Fermi then suddenly burst out, saying "Where is everybody?" to general laughter, as they all instantly understood that he was talking about extraterrestrials.

The premise of Fermi's paradox is the following. The Milky Way is about 13 billion years old and contains between 200 and 400 billion stars. Our own sun is only a little over 4 billion years old, suggesting that there have been stars around for a very long time. If Aliens are common in the galaxy, there has been plenty of time for them to have evolved—perhaps hundreds of millions of years or more before humanity—and have visited Earth. So where are they?

While Fermi's outburst is the origin of the paradox, the question was re-visited in 1975 by Michael Hart (leading some to call this the Fermi-Hart paradox). Hart published "An Explanation for the Absence of Extraterrestrial Life on Earth" in the *Quarterly Journal of the Royal Astronomical Society*. In this article, he explored some of the reasons why we hadn't been contacted yet, from reasons of simple disinterest of the Aliens to either colonize the galaxy or to contact us to the idea that the Earth is being treated as a nature preserve. Perhaps some form of *Star Trek's* Prime Directive applies, whereby civilizations are not contacted until they develop the capability for interstellar travel. As we recall from chapters 3 and 4, these kinds of explanations were offered in *The Day the Earth Stood Still* and, of course, *Star Trek*. What Hart was able to show was that technology wasn't the problem. Taking some simple assumptions, Hart showed that a civilization that sent out two craft traveling at 10% of the speed of light to nearby stars and then spent a few hundred years developing infrastructure to build another pair of slow-moving starships could completely populate the Milky Way in just a couple of million years. Given the timescales involved, from the fact that there are stars that are billions of years older than the sun, it seems impossible that we should not have been visited before. If intelligent extraterrestrial life is even slightly common in the galaxy and only a few species have mankind's curiosity and exploratory nature, it seems that we would know by now that we are not alone. Hart concluded that it was a distinct possibility that mankind might well be one of the earliest-developing intelligent species in the galaxy. In short, *The X-Files* tagline "We are not alone" could well be gravely incorrect.

Of course, the answer to the question is unknown and hence the reason why the term "paradox" is applied to it. Author Steven Webb explored the question in his delightful 2002 book *If the Universe Is Teeming with Aliens, Where Is Everybody? Fifty Solutions to Fermi's Paradox and the Problem of Extraterrestrial Life*. Peter Ward and Donald Brownlee's 2003 book *Rare Earth: Why Complex Life Is Uncommon in the Universe* is equally enjoyable, and this book takes the position that it is difficult for a planet to develop intelligent life. The book describes the many ways in which planetary disaster can interrupt the development of sentient life on a planet.

No matter how carefully thought out, arguments of the sorts advanced in these books and others like them must defer to data. And to figure out what sorts of data are needed, it is helpful to have a guiding paradigm. The question of nearby extraterrestrial life has long been guided by a simple equation developed in 1961.

The Drake Equation

Frank Drake has been a leader in the field of searching for extraterrestrials for more than half a century. He started out as a radio astronomer using the National Radio Astronomy Observatory, located in Green Bank, West Virginia. While he did research into the physics of radio-emitting astronomical objects, he is most known for using his radio expertise to search for civilizations living on planets around nearby stars. The idea is quite simple. Humanity has been emitting radio or television broadcasts for about a hundred years. Since radio travels at the speed of light, this means that for a hundred light-years in all directions, a sufficiently advanced civilization might hear our broadcasts and know we're there. Turning that logic around, we can instead listen to the cosmos using our own radio telescopes. If there are nearby civilizations with a comparable level of technology, it seems like we should be able to hear them. Since we don't know if it will ever be possible to travel at speeds faster than light, perhaps the fastest way to communicate between stars will be radio communication. If there is a vast galactic civilization out there, perhaps we could intercept broadcasts from one star to another. Given that we have the right kind of equipment, we really should listen. Perhaps our first communication with an extraterrestrial civilization will be when we pick up a stray broadcast of an Alien version of *Gilligan's Island*.

We will talk more about the history of searching for extraterrestrial life by way of radio in a short while, but the current subject is the Drake equation. In 1960, Drake had undertaken his first radio survey of nearby stars and had been asked by the National Academy of Sciences to convene a conference on the question of extraterrestrial life. This conference was held at Green Bank in 1961. Drake realized that he needed an agenda for the conference, so he put together his famed equation as a way to guide the discussion. He wrote down all the things you need to know in order to predict the number of extraterrestrial civilizations in the galaxy. As we will see, this equation doesn't stem from any particular theory of life formation but is essentially a classic Fermi problem. Because of the nature of the conference, it focused on the chances of receiving an extraterrestrial radio transmission.

The Drake equation is very simple

$$N = R_* \times f_p \times n_e \times f_l \times f_i \times f_c \times L$$

where:

$N =$ the number of civilizations in our galaxy from whom we might receive a radio broadcast

$R_* =$ the average rate of star formation in the galaxy

$f_p =$ the fraction of stars with planets

$n_e =$ the average number of planets that can support life around those stars with planets

$f_l =$ the fraction of those planets with habitable planets that develop life

$f_i =$ the fraction of those planets with life that develop intelligent life

$f_c =$ the fraction of stars with intelligent life that develop radio or other detectable technology, and

$L =$ the period of time during which the civilization will be detectable.

For a scientist, there are many potential criticisms one can level at the Drake equation. Obviously the relevant rate of star formation isn't the rate we see now, but rather the rate several billion years ago. Further, the equation predicts the number of spontaneously arising civilizations. A weakness of this approach is that if some culture did develop interstellar travel, the equation quite ignores the creep of culture across the galaxy. For instance, even if the creation of intelligent life is exceedingly rare—so rare that only one culture developed our level of technology before we did—if they had developed the ability to travel interstellar distances and had humanity's wanderlust, this could result in far more radio-emitting planets than the Drake equation suggests.

Still, the equation gives us a starting place and tells us the kinds of parameters that are important for researchers to consider as they try to figure out a reasonable guess for the number of technology-using extraterrestrial civilizations in the galaxy. It is perhaps obvious that most of these factors are not known at all, although we are not completely uninformed as to what constitutes a reasonable range of values.

We can take a look at the modern realistic guesses. To begin with, the rate of star formation (R_*) is reasonably well determined from astronomical research. In addition, we are now able to estimate the fraction of stars around

TABLE 7.1. Values for the Drake equation showing different scenarios

Equation	R* (/year)	f_p	n_e	f_l	f_i	f_c	L (years)	N
							Element	
Drake (1961)	10	0.5	2	1	0.01	0.01	10,000	10
Modern (optimist)	20	0.5	2	1	0.1	0.1	100,000	20,000
Modern (pessimist)	7	0.5	0.01	0.13	0.001	0.01	1,000	0.00005
Modern (realist)	7	0.4	2	0.33	0.01	0.01	10,000	1.8

Note: Values for the Drake equation are essentially educated guesses and reveal prejudices of the person making the assumptions. The optimist scenario is rendered implausible by modern measurements, at least in our nearby galactic neighborhood. The pessimist scenario ensures that we are essentially alone in the galaxy. The realist scenario suggests that there are but a handful of technologically advanced civilizations in the galaxy, although there are perhaps two hundred intelligent species who have not yet developed detectable technology.

$R*$ = average rate of star formation; f_p = fraction of stars with planets; n_e = average number of stars that can support life on these planets; f_l = fraction of those planets that actually develop life; f_i = fraction of these planets that develop intelligent life; f_c = fraction of these planets that develop radio technology or method of communication; L = period of time that civilization will be detectable; N = number of civilizations from whom we might receive a radio or other broadcast.

which planets form (f_p). We will discuss these studies more a little later, but planet-hunting research leads us to believe that about half of stars develop some sort of planetary system. Our current technology preferentially finds systems in which a large planet orbits close to the star, and we find that something like 40% of the surveyed stars in our stellar neighborhood have this property. When we combine this observation with the fact that there are no doubt stars that have planets but without a huge planet like Jupiter, the actual fraction of planet-hosting stars is no doubt somewhat larger.

Determining the number of habitable planets (or moons of gas giant planets) given that a planetary system has formed (n_e) is much harder. Investigating this question is a hot research topic in the scientific community, with the launch in 2009 of the Kepler mission. By the time you read this book, what you learn here will definitely be out of date. You should keep your eye on this quickly evolving topic.

We don't know the fraction of planets that are potentially habitable that actually host life (f_l), but it would seem that this should be fairly high. The fact that life evolved on Earth so quickly after the planet cooled enough to allow liquid water suggests that life might develop easily. One can use statistical techniques to use the period of time it took life to develop on Earth to come up with a lower limit on this fraction. Unless there are exceptional qualities of Earth that make it a nonrepresentative planet, it seems that the probability of an Earth-like planet developing life is higher than 20%.

Another poorly known factor is the fraction of planets that develop life that go on to evolve intelligent life (f_i). There are two very distinct ways to think about this. The first school of thought suggests that the development of intelligence is inevitable. Proponents point to the observation of a steady increase in the intelligence of species over the course of the eons. Under this way of thinking, the formation of intelligence is approximately inevitable, given enough time. A contrasting way of thinking points to the fact that there have been millions of vertebrate species, and only humans have developed the kind of intelligence we have. If something had caused the hominid line to go extinct 100,000 years ago, there is no indication that another species would have gained intelligence in the intervening time. Another factoid is that, even though dinosaurs ruled the planet for about 150 million years, there is no evidence of significant intelligence developing during that time. This suggests that the development of intelligence is rather rare.

The fraction of intelligent and technologically advanced civilizations that will announce their presence (f_c) would seem to be rather high. We have only us to use as an example. We rarely intentionally attempt to communicate with potential civilizations around adjacent stars, but we don't have to do this intentionally. After all, since the early days of the twentieth century, mankind has been broadcasting its existence into the cosmos. Figure 7.1 shows us the radio and television bubble surrounding the Earth as of 2010.

The final factor, which is the lifetime that a civilization will emit a detectable signal (L) is also not well known. Civilizations on Earth tend to be at their peak for several hundred years, but subsequent civilizations often use technology from the antecedent civilization. Further, we don't know for how long we will use radio and broadcast television to communicate. However, unless war decimates human life on Earth, either from atomic cataclysm, extreme and rapid environmental devastation or some sort of intentional biological warfare, it seems probable that ongoing use of radio, television, or some sort of electromagnetic emission is likely to continue for hundreds, if not thousands, of years.

Given the difficulty inherent in determining the parameters that go into the Drake equation (and even whether the equation is an appropriate mathematical representation of the question), it is inevitable that there will be ongoing uncertainty as to the number of technologically advanced civilizations

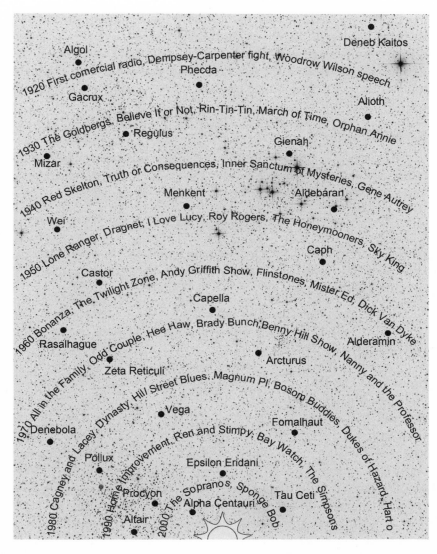

Deneb Kaitos
Algol
1920 First comercial radio, Dempsey-Carpenter fight, Woodrow Wilson speech
Phecda
Gacrux
Alioth
1930 The Goldbergs, Believe It or Not, Rin-Tin-Tin, March of Time, Orphan Annie
Regulus
Gienah
Mizar
1940 Red Skelton, Truth or Consequences, Inner Sanctum of Mysteries, Gene Autrey
Menkent
Aldebaran
Wei
1950 Lone Ranger, Dragnet, I Love Lucy, Roy Rogers, The Honeymooners, Sky King
Caph
Castor
1960 Bonanza, The, Twilight Zone, Andy Griffith Show, Flinstones, Mister Ed, Dick Van Dyke
Capella
Rasalhague
Alderamin
1970 All in the Family, Odd Couple, Hee Haw, Brady Bunch, Benny Hill Show, Nanny and the Professor
Zeta Reticuli
Arcturus
Vega
1970 Cagney and Lacey, Dynasty, Hill Street Blues, Magnum PI, Bosom Buddies, Dukes of Hazard
Denebola
Fomalhaut
Pollux
Epsilon Eridani
1990 Home Improvement, Ren and Stimpy, Bay Watch, The Simpsons
Procyon
2000 The Sopranos, Sponge Bob
Tau Ceti
Alpha Centauri
Altair
Hart o

FIGURE 7.1. Nearby stars have been graced by our radio and television signals for nearly a century. It is easy to imagine that our first contact with Aliens will not be intentional but rather by an extraterrestrial civilization intercepting reruns of *Ren and Stimpy*. It's kind of a sobering thought.

we expect in our galaxy. Drake's equation clearly assumes that the civilizations are independent, with no cross-pollination. It also doesn't allow for a civilization that spans a large segment of the galaxy. A dispersed civilization might allow for bits and pieces of the society to go extinct, but it is more difficult to believe a thriving culture spanning millions of star systems would disappear entirely.

Kardashev Scale

In 1964, Nikolai Kardashev formalized the idea of variations in the level of technological achievement of extraterrestrial civilizations. He defined three distinct classes.

Level I: A civilization that can totally utilize all the energy from a star that reaches a planet
Level II: A civilization that can totally utilize the energy resources of a star
Level III: A civilization that can totally utilize the energy resources of an entire galaxy

Subsequent extensions of Level IV (utilizing the energy output of the visible universe) and Level V (utilizing the energy of the multiverse) are later modifications and rarely used.

It is perhaps obvious that a Level III civilization will be more detectable than a Level I civilization, in the same way that a spotlight is easier to view from great distances than a candle. As we read further into the searches for extraterrestrial life, we must keep in mind the fact that when we look beyond our solar system, we're not necessarily looking for life with the same technological level as our own. It is quite possible that an extraterrestrial civilization might have a significant head start on us. To a degree, the current technological phase (i.e., the phase in which both electricity and radio have been mastered) of our civilization is only about 100 years old. Imagine the kinds of technology we might master by the year 3,000. Just a mere millennia is likely to bring us unfathomable advances. Now imagine that a civilization in our stellar neighborhood hit our level of technological development when the Neanderthals were dying out, when a lineage of Miocene ape experienced the mutations that lead to *Homo sapiens*, or even when the impact at Chicxulub killed the dinosaurs. Those Aliens would presumably have mastered technologies of which we can only dream (or, more likely, beyond anything we can dream of). Given the raw numbers of stars out there and working under the assumption that the Earth is not an exceptional planet, it seems inevitable that any intelligent extraterrestrial species we encounter will be more technologically advanced than us. So, what do we see?

The Big Ear

The thought of using radio to listen for life on other planets is an old one that can be traced back at least as far as Nikola Tesla. In Colorado Springs in 1899, he believed that he had perhaps established communication with extraterrestrials, although he was uncertain whether it was from Mars or Venus. (Keep in mind that this was at the height of the media frenzy about the question of canals on Mars.) He received in his equipment groups of clicks of one, two, three, or four. This was reminiscent of how the Martians communicated in the 1952 movie *The Red Planet Mars* (discussed in chapter 3). He wrote of the experience in the February 19, 1901, issue of *Collier's Weekly* (as well as many other places; Tesla was both a technical genius and a prolific popularizer). He said, "There would be no insurmountable obstacle in constructing a machine capable of conveying a message to Mars, nor would there be any great difficulty in recording signals transmitted to us by the inhabitants of that planet." His work in this area has long since been discredited, with many suggested explanations, the most likely of which is that he simply didn't understand his equipment. This isn't incredibly surprising, as Tesla's pronouncements were often more spectacular than his accomplishments, and his accomplishments were very spectacular indeed. The most important point is that the idea of using radio to communicate to other planets has its antecedents in the very beginnings of mankind's use of the technology.

Although Tesla's efforts were perhaps the first, he was not alone. About two decades later, Guglielmo Marconi made similar claims. Marconi and Tesla were favorites of the media (think Steve Jobs in an era where this kind of technological innovation was rare) and received substantial attention in the press. In 1919, Marconi believed it possible that he received radio broadcasts from beyond the Earth. His evidence included simultaneous reception of signals in New York and in London, suggesting that the source was not local. Critics pointed out that radio receivers at the Eiffel Tower and in Washington, D.C., heard nothing. The *New York Times* had multi-week coverage of the story, often on the first page and above the fold. The editors of the *New York Times* suggested that perhaps it would be better if mankind did not contact life on other planets. Their reasoning was that the other life, being older and thus more advanced, would have technology far beyond ours and that mankind was not ready for it. This caution was seconded much later in the twentieth century by physicist Stephen Hawking, who pointed out that when an advanced culture encountered a less advanced one, the less advanced culture invariably suffered. This is another reason to think it wise to "lay low." In retrospect, both Marconi and Tesla were monitoring frequencies that were too low to penetrate the Earth's ionosphere, but they were still efforts that electrified the public.

The periodical *Scientific American* was just as cutting edge a magazine in 1919 as it is today and, within a couple of weeks, they had penned an article on Marconi's claims, followed a couple of months later by a truly forward-thinking article. Marconi was talking about having received an occasional letter of Morse code, and *Scientific American* pointed out the inherent difficulties of using such a code for interplanetary communication. They went so far as to advance a way to communicate with Mars that might work, anticipating by decades a similar message broadcast from the Arecibo radio telescope in Puerto Rico. In this much later attempt, mankind intentionally beamed a signal into space with the hopes that it might one day be intercepted by extraterrestrials. *Scientific American's* 1920 proposed message is shown in figure 7.2.

While the belief in Martian canals had disappeared among most scientists with the 1909 Martian opposition, the idea lived on in the public imagination for much longer. During the 1924 Martian opposition, in which Mars and Earth were particularly close, another attempt was made to search for radio signals from our neighbor planet. On August 21 to 23 (the date of the opposition), the United States declared a "National Radio Silence Day," which was somewhat misnamed. What was actually advocated was that all radio traffic was turned off for five minutes, every hour, on the hour, over a 36 hour period. During that time, receivers were to listen to the heavens, looking for that Martian signal. The U.S. government got into the act with the chief signal officer of the army telling his radio stations to be vigilant for unusual transmissions, while the secretary of the navy directed the most powerful radio stations under his command to broadcast minimally and keep an ear out. Very few of the commercial stations complied, except one in Washington, D.C. The attempt was a dismal failure, scientifically speaking, but it was an interesting idea.

Over the course of the next few decades, the idea of interplanetary communication persisted among a few, including amateur radio operators. The simple fact was that the technology of the era was not really up to the project. Further, after 1930 or so, the scientific community had essentially dismissed the possibility of intelligent life on Mars, which meant the new goal was in-

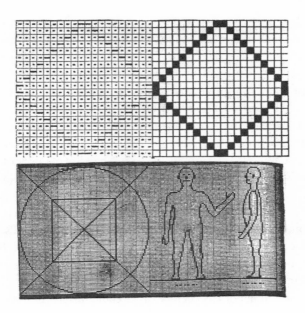

FIGURE 7.2. This figure from the March 20, 1920, issue *of Scientific American* shows an early attempt to design a message that would be understandable by an extraterrestrial civilization. It is hard enough for an English-only speaker to write a message that would be understood by (say) a literate person who reads only Chinese, let alone a culture that is as different from humanity as an Alien civilization is likely to be. *Scientific American.*

terstellar communication. This capability was definitely beyond the capability of the equipment of the time.

In the 1950s, the field of radio astronomy was born. Astronomers knew that astronomical bodies would emit electromagnetic radiation beyond the visible spectrum. Large radio dishes began to be built to study things like the galactic center, the sun, and similar sources. And this is where we again meet Frank Drake (of the Drake equation).

Frank Drake was a radio astronomer working to build the new 140 foot radio telescope in Green Bank, West Virginia. This huge antenna was housed at the National Radio Astronomy Observatory (NRAO) and signaled that a national decision had been made on whether it was better for a country to fund a large, central government laboratory or many, smaller research efforts, dispersed across the various universities and where individual researchers could exercise greater control over their research interests. Big won.

Drake long had an interest in searching the heavens for extraterrestrial radio signals. During some earlier astronomical research, he had picked up a transient signal that was never explained. While he was not prone to out-landish claims, the thought crossed his mind that perhaps the signal was not emitted from any transmitter on Earth. He tabled this idea while he pur-sued a more traditional research career. The quality of his research was what smoothed his way to a position at the newest and biggest radio facility around.

NRAO was a lively place, with groundbreaking in 1957 and construction on the 140 foot telescope beginning in 1958. A handful of physicists and buck-ets of money were tasked with building a breathtaking new radio astronomy laboratory. The big antenna was much bigger than earlier antennas, and its construction was plagued with more than the usual problems that come from building something never built before. Thus, in the meantime, NRAO decided to construct an 85 foot telescope, as this was much less of a technical chal-lenge and would get the facility off and running. The 85 foot telescope started operations in early 1959.

It was the summer of 1959 before Drake could turn to his idea of using the NRAO facility to search for extraterrestrial signals. By consensus, the scien-tific staff agreed that (1) first priority had to go toward more traditional radio astronomy research and (2), given the potentially sensational nature of the extraterrestrial search, they should keep their efforts quiet. The idea was to do a simple search and to see what they could see without fear of interference that might accompany a negative newspaper story.

However, the world of high end scientific research was no less cutthroat in 1959 than it is today. There is a large degree to which it is true that in high stakes science, there is first and there is not-first. There is no second. In Sep-tember 1959, a paper was published by two theoretically inclined physicists, Philip Morrison and Giuseppe Cocconi, which discussed how one might go about doing a radio search for extraterrestrials. Worried about losing credit for what was undoubtedly an important bit of research, Otto Struve, director of the Green Bank facility, announced the effort in a series of lectures given in November at Massachusetts Institute of Technology—just a mere two months after the theoretical paper came out. The lectures brought with them instant attention in the press. Articles in *Time* magazine, the *New York Times*, and the *Saturday Review* told the public that astronomers were going to try to listen for Alien broadcasts. The *Saturday Review* reported on Struve's presentation: "He was struggling against going too far without falling too short of expressing what many astronomers have come to believe—that other intelligent beings share our occupancy of the cosmos, that some of them are very probably superior to us, culturally, and that our existence is suspected by if not definitely known to them."

The response in the press was typically positive. The exposure brought with it donations of cutting-edge amplifiers, which would enhance the performance of the equipment. The era of modern searches for extraterrestrial broadcasts had begun.

The term SETI (Searches for Extraterrestrial Intelligence) was not coined until the mid-1970s, but that is exactly what Drake and colleagues were doing. Drake named the 1960 effort "Ozma" after Princess Ozma in L. Frank Baum's sequels to *The Wizard of Oz*. Baum claimed to be in radio contact with Oz, which is how he learned of his stories. Drake and company were attempting to contact a land far weirder than Baum's fictional kingdom.

On April 8, 1960, Operation Ozma began operations. The team looked at two stars, Tau Ceti and Epsilon Eridani. Both were thought to be sufficiently like our sun to be interesting. Subsequent analysis has tempered somewhat the enthusiasm for these stars, but they remain targets for modern planet-hunting attempts. While observing these stars, there appeared to have been a transient signal from Epsilon Eridani, although this signal turned out to have a terrestrial origin. The study was a first try and therefore covered a limited amount of the radio spectrum. No extraterrestrial signal was observed.

This brings up an important point. The radio spectrum is quite broad and artificial broadcasts are quite narrow. The Federal Communications Commission has allocated the range from 9 kHz to 275 GHz for use. Translated to wavelength, these radio frequencies range from a fraction of an inch to a few miles in length. Although we shouldn't expect extraterrestrials to comply with human choices, a single AM radio station can occupy about 20 kHz of that range, while an FM station can occupy 200 kHz. So, in the semi-arbitrary range used by humans, one can fit about 13 million AM stations and over a million FM ones. In order to see the technical details of a broadcast, the spectrum must be split even finer still. As we will see below, SETI searchers generally concentrate on a fraction of the possible radio spectrum, but still end up needing to simultaneously search hundreds of millions of radio channels.

While the radio spectrum is broad, the range of frequencies scientists use for SETI studies has historically been narrower. The particular range that is used has been selected to avoid frequency ranges that are noisy from naturally occurring sources. For instance, the Earth's atmosphere radiates copiously for wavelengths below about an inch, while the galaxy radiates for wavelengths above about a foot. Although these thresholds aren't perfectly sharp, SETI scientists listening outside this wavelength range will have to contend with a much louder "radio hiss." Further, we should remember that NRAO was really a radio astronomy facility and not a SETI one. This is not unusual and, even today, when a new facility is built, SETI is almost always a secondary consideration. Luckily the limitations imposed by unwanted radio noise affect radio astronomers and SETI researchers equally, allowing the same equipment to be used for both goals. NRAO was funded to study astronomical phenomena and so the equipment was optimized for a radio wavelength of about 8 inches, as this would allow researchers to study interstellar hydrogen to look for magnetic fields.

As we have seen, Ozma failed to observe any SETI signal, but it generated great excitement, culminating in the November 1961 conference at which the Drake equation was unveiled. A new era in exploratory science had begun.

In the intervening 50 years, there have been many SETI efforts, although there have certainly been gaps during which no observations were attempted. One telescope was built in Delaware, Ohio, and began operations in 1963. Funded by the National Science Foundation and operated by the Ohio State University, the facility was called Big Ear. From about 1963 to 1971, the facility was used for traditional radio astronomy research, mapping extrasolar radio sources. However, after NSF funding was cut, the facility turned to SETI research, operating from 1973 to 1995. In 1977, the so-called Wow! signal was reported, named for a prominent "Wow!" written on the printout where the signal was first observed (figure 7.3). It is still considered to be the most interesting extrasolar candidate radio transmission recorded so far (which doesn't mean it was really extrasolar in origin). From the telescope's orientation at the time, the signal appears to have originated in the constellation Sagittarius, near the Chi Sagittarii star group. Despite many additional attempts to observe this region of the sky, no similar signals have ever been observed.

As discussed above, in order to find radio transmissions over a very narrow wavelength range, it is important to be able to split up the radio spectrum very finely. The 1980s ushered in an era where it was possible to simultaneously study a million radio channels, followed by the billion-channel era of the 1990s. Because of these technical improvements, the rate of progress has increased rapidly, much like computer technology grows by leaps and bounds. Originally SETI searches were funded by the U.S. government, but they were always vulnerable to ridicule by budget-conscious politicians who could make some political hay by denigrating searches for "little green men." In 1983,

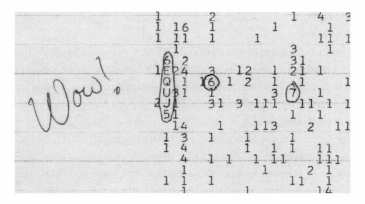

FIGURE 7.3. The "Wow!" signal was recorded by a SETI researcher at the Big Ear radio astronomy facility. The Ohio State University Radio Observatory and the North American AstroPhysical Observatory (NAAPO).

government funding was finally cut. SETI advocates persevered even without that source of money and in 1984, the SETI Institute began operations as a nonprofit, backed by private funding. First observations began in 1992.

Current SETI searches are dominated by the Allen Telescope Array, named after Paul Allen, the project's benefactor and cofounder of Microsoft. The technical effort was initially helmed as a collaborative effort between the SETI Institute and the University of California, Berkeley, although Berkeley has since pulled out and transferred the facility to SRI International. Even with Allen's generous donations, the facility requires additional funding to successfully operate. Budgetary shortfalls forced the facility to go into mothballs in April 2011, although sufficient monies were procured to resume operations in December 2011. As of this writing, continuing operations remain in doubt. Given the inarguable consequences of a successful measurement of detection of a SETI signal, and the very modest needs, it seems to me that this is an unconscionable lapse in research priorities. The costs are small, and the potential payoff is incalculably huge.

So, what is the status of SETI searches in 2012? Well, so far, we haven't found a radio signal originating from extraterrestrial intelligence or, if we have, we haven't recognized it. We should also dispense with the conspiracy theories suggesting that the government is in contact with Aliens and just hasn't told us. The big SETI efforts are civilian-run and further run by people with a lifelong passion for searching for our interstellar neighbors. In a world of blogs and leaks and rumors, I find it utterly inconceivable that a secret of this magnitude could successfully be hidden. No ET signal is being covered up by the government.

But what do we know that we didn't know 50 years ago? Well, the first thing we know is that there aren't many radio-emitting civilizations similar to ours currently living in our local stellar community. The hopes of a universe filled with neighbors much like us have not proven to be true. As much as it makes me unspeakably sad, we don't live in the *Star Trek* universe.

However, no matter how easy it is for SETI opponents to point at the failure after half a century of effort, SETI advocates can point at many possible explanations of why we've not yet succeeded. We have restricted the bulk of our studies to a limited range of radio space. Perhaps the Aliens have elected to broadcast in a different range. Perhaps the Alien's signals haven't reached us yet. Indeed, in the movie *Contact*, Aliens located near the star Vega first learned about Earth when the 1932 broadcast of the Olympics reached them. The Aliens recorded the broadcast and sent it back to us, amplified greatly. In their transmission, they interspersed their own message.

While we have no idea how we will one day encounter an extraterrestrial radio broadcast (if ever), under that plausible scenario, maybe the broadcast just hasn't arrived yet. If Aliens living under the sun of Aldeberan (65 light-years away) received the 1932 broadcast and replied instantly, we wouldn't hear their response until 2062. (Aldeberan, being a red giant, is an unlikely place to find an indigenous extraterrestrial civilization, although it is obviously possible that Aliens could have travelled there, so it could host a broadcasting antenna.)

While SETI advocates quite rightfully remind us that there are many perfectly reasonable reasons why we have not heard a radio broadcast by Aliens and we should continue looking, it is safe to say that the data taken thus far can rule out a nearby Kardashev level II or III civilization. It also seems equally safe to say that we probably don't have a neighbor in our stellar neighborhood who has been broadcasting radio for hundreds of years. Nearby intelligent life, at least of the radio-broadcasting variety, seems to be rare. But the galaxy is big, and there is no reason to give up just yet.

Where Could Aliens Be?

If nearby intelligent and technologically advanced life is rare, why is that so? The Drake equation, for all its imperfections, tells us what parameters are the most important. We know that the universe makes stars and further we know it makes planets. As of this writing (spring 2012), NASA's Kepler spacecraft has observed 2,321 planets orbiting distant stars. In December 2011, NASA announced the first observation of a planet that circles a distant star in the "habitable zone," which means that the planet could contain liquid water. That planet is called "Kepler-22b" and is the first of no doubt many such observations. By the time you read this, these numbers will be terribly out of date. Already, the Kepler team has announced another fifty candidates of potentially habitable extrasolar planets that need further study to be sure that they're real.

The most important fact is that scientists no longer need to speculate about planets around other stars. We're observing them directly. The Kepler team's best estimate is that at least 5% of stars include at least one Earth-sized planet and at least 20% of the stars have multiple planets. Given the youth of this field of research, it is highly likely that, as the sensitivity of the equipment improves, we will find the actual numbers are even higher. The Kepler spacecraft is observing about 150,000 of the approximately 300,000,000,000 stars in the galaxy. This is an absolutely fascinating time to be an extrasolar planetary astronomer, and the fun is only beginning.

If there are many stars and many planets, then the next question is how many of those planets harbor life and how many of the planets with life host intelligent life? These numbers are much harder to estimate, but they remain the crux of the question.

In order for Aliens to exist, they need a stable environment. Life on Earth developed several billion years ago. Complex animal life was first preserved in the fossil record about 530 million years ago. Mammals appeared about 210 million years ago, and the first primates originated perhaps 50 million years ago. Finally, the first hominids originated about 17 million years ago and our own species, *Homo sapiens*, is only about 50,000 to 100,000 years old.

It took billions of years for intelligence to develop on Earth. If, at any time during those long eons, the Earth became uninhabitable for life like us, we wouldn't be here. This doesn't mean that the climate on the planet must be stable, after all there have been periods during which the entire Earth was frozen over, and huge volcanic eruptions and meteor and comet strikes have killed off vast swathes of species. But there have been no "sterilizing events."

What might constitute a sterilizing event? Well, one theory of the origin of Earth's moon is that a Mars-sized planetoid collided with an early version of Earth perhaps 4.5 billion years ago. This collision would have thoroughly melted any crust that had formed by that time. An impact like that would have extinguished life.

Another danger to a planet's biosphere would be a nearby supernova. If a supernova occurs within a few tens of light-years from the Earth, it could deplete a large fraction of the Earth's ozone. Since ozone protects the Earth from the sterilizing ultraviolet light from the sun, loss of a large fraction of the Earth's ozone would be a catastrophic event.

Even more dangerous (although much rarer) is a gamma ray burst. Gamma ray bursts are a special class of supernovae that occur when a very massive, rapidly rotating star explodes. Rather than having the energy expand in a spherical pattern, the energy is blasted into two beams shooting out from the poles of the star. To give a sense of the amount of energy we're talking about, a gamma ray burst can be observed billions of light-years away. The energy of a typical burst releases in just a few scant seconds as much energy as our sun will release over its entire ten billion year lifetime. The energy release of a gamma ray burst is an astoundingly dangerous occurrence. Luckily, they are rare. One occurs perhaps every 100,000 or 1,000,000 years in a galaxy the size of the Milky Way, and they are a danger only if the beams are pointed directly at us. The nearest candidate for a gamma ray burst is one of the two stars in the binary star system WR 104. It is located about 8,000 light-years away from us in the general direction of the galactic center and its axis appears to be pointed roughly in our general direction. The chances that the burst is pointed exactly at us are rather small, so there is no reason to worry. But if it were, it could badly damage the ozone layer and thus devastate the biosphere.

Dramatic events like supernovae and gamma ray bursts are not required to seriously damage a planet. Little things like the evolution of a star's output over the course its lifetime can also be the source of destruction of life. Around any star, there is a range of distances in which water can be liquid. Estimates vary, but the current habitable range for our sun is about 0.97 to 1.37 times the Earth's orbit. Thus the Earth is barely inside the habitable zone. Were the radius of the Earth's orbit just 10% smaller, Earth would be too hot for life.

It's actually more complicated than that. The energy output of the sun has evolved over time. Several billion years ago, it is thought that the energy output of the sun was about 80% of what it is now. That suggests that the habitable zone for the solar system would exist for smaller radii than today. Back then, the minimum and maximum habitable distances from the sun was 0.80 and 1.15 times the radius of the Earth's orbit, respectively.

What matters is the "continuously habitable zone," which is the biggest minimum and the smallest maximum radius over the course of the sun's lifetime within which life can exist. So far, the continuously habitable zone is the very narrow region of about 0.97 to 1.15 times the radius of the Earth's orbit. A planet outside that small region would not remain habitable long enough for intelligent life to develop.

You should be aware that the numbers offered here are actually quite contentious within the astrobiology community. Different experts have come up with different estimates. Considerations like the chemical makeup of the atmosphere and effects arising from plate tectonics can change the outcome of the calculations. Further, the discussion here is concerned with energy from the central star. There are other energy sources, like the tidal flexing of moons in close orbit to a big planet. This is the reason that moons like Jupiter's Europa are considered to be candidates for places where life might have arisen.

But the general idea is still largely valid. For life like us, which depend on the warmth of our sun to survive, there is a "Goldilocks Zone" around the central star—not too warm and not too cold. Other stars are hotter or colder and the details of the habitable zone will adjust accordingly. But still, the star's energy output will have to be stable enough so that a planet on which life begins will continue to be a hospitable spot for life to evolve.

Even if a star is very stable, it is equally imperative that the orbit of the planet be stable and pretty circular. A highly elliptical orbit will bring the planet alternatively too close and too far from the central star. While the orbits of the planets of the solar system are elliptical, they are close enough to being circular that the difference is indistinguishable to the naked eye. A slightly elliptical orbit is allowed, as long as the orbit doesn't leave the habitable zone.

Further, it's not enough that the orbit of the planet hosting life be nearly circular. If another planet in the planetary system has an eccentric orbit, the gravity of the uncooperative planet could eject the hospitable planet either into the star or out into the coldness of interstellar space.

There is a very long list of things that have to go right to have a planet that can (1) have life begin on it and (2) allow life to persist long enough to evolve an inquisitive intelligence like our own. It may well be that this is an extremely rare situation.

Wrap Up

If you look at the simplest of facts, for instance Carl Sagan's oft-quoted estimate of there being "billions and billions" of stars in the universe, including that there are some 300 billion stars in our galaxy alone, it seems unfathomable that there is not life elsewhere in the universe. If not, to repeat another commonly made statement, it sure seems like an awful waste of space.

The history of science has been an unremitting onslaught of the mediocrity principle. While humanity once thought that the Earth occupied a special spot on the solar system and then the cosmos, we now know that in many respects, the Earth is a small planet around an undistinguished star, orbiting in an undistinguished location in an undistinguished galaxy. Mankind was once thought to be a species of an entirely different kind, given dominion over every living being that moved on the Earth. We now know that mankind is instead a single species, with a genetic heritage shared by all other organisms on the planet.

If Earth and mankind are, indeed, entirely ordinary, it seems inevitable that there must be life on other planets; that we are one day fated to encounter species like us in many ways, driven by the instincts to reproduce and survive, and no doubt utterly alien in form and thinking.

Yet our initial searches for stellar neighbors have come up short. Despite diligent and imaginative attempts to listen in on conversations of our fellow travelers, we have no evidence that there is anyone out there.

This is an interesting time in the field of astrobiology. While the funding for direct SETI searches is less stable than it should be, the planet finders are going like gangbusters. Planets are being found every day. The technology and techniques have improved to the point of being able to find Earth-like planets within the next couple of years. Techniques for looking at the atmospheres of extrasolar planets are conceivable. It may well be that the next few years or decades will answer the question "Are we alone?" once and for all.

THE VISITORS

Kindly take us to your President.

Alex Graham, in a March 21, 1953, *New Yorker* cartoon in
which two Aliens are talking to a horse.

Imagine someday that an amateur astronomer is taking pictures of the sky
and finds that one of the thousands of dots on the screen has moved. After he
checks the online ephemerides, he reports that a new comet has been found.
The professionals turn their bigger telescopes to the comet candidate, and
they project its path and find that it will come close enough to the Earth to
make them nervous. Around the planet, world leaders are notified of a comet
that will have a near Earth encounter. Concerned leaders are reassured that
space is large and the Earth is small. The precision of the predictions for the
comet's path isn't great and it will probably be a near miss; a spectacular lightshow to be sure, but a miss nonetheless.

Subsequent study reveals that the comet's path really does seem to be
intersecting the Earth's. Perhaps the Earth really does have a big bulls-eye
painted on it. Discussions in the halls of power revolve around whether the
public should be notified or whether quiet preparations should be made to

ensure the survival of civilization if, as happened once 65 million years ago, a comet or even a piece of the comet plummeted out of the sky, striking the Earth and causing enough damage to drive species to extinction.

This being the twenty-first century, secrets are impossible to hide and a Facebook or a blog post is made, the press gets wind of the story, and people are told. Like the Y2K case, the press will write hysterical stories. Survivalist and religious organizations get a spike in membership, while some people discount the situation as typical media hype.

Astronomers across the world keep round-the-clock watch on the comet, firming up the predictions of the trajectory. There is no longer any doubt. The comet is aimed directly at Earth. There's only one problem. As it gets near, the object appears to be slowing. This can't be explained by orbital dynamics and the fact perplexes physicists. To the UFO community, the message is clear. The comet isn't a comet at all, it's an Alien spaceship. While the claim seems pretty ludicrous, scientists admit that the claim would explain things. By now the comet (or spacecraft) has come under observation by amateur telescopes and its size is known. It's pretty big, whatever it is.

As astronomers watch, the slowing object misses the Earth and settles into a high orbit. This behavior answers the question. It's not a comet or an asteroid, but rather some phenomenon under intelligent control.

As the radar watches, a small object detaches itself from the bigger one and approaches the Earth, descending slowly. The object enters the atmosphere and appears to be headed to Washington, D.C. The combat air patrol permanently stationed over the city since September of 2001 heads to intercept, while fighters scramble from nearby airfields to lend support. The commander of the air force asks the president for orders. She's a tough old bird, so she orders the vice president onto Air Force 2, tells the air force to hold their fire and waits in the Oval Office. As the fighters converge on the descending object, they see it is egg-shaped, with the wide end appearing to be the front.

Surrounded by dozens of air force fighters, the unknown craft descends onto the Mall in front of the White House, landing on the Ellipse. The military had earlier dispatched quick response troops, which now surround the egg. Secret Service agents on the top of the White House aim Stinger missiles at it, while the president watches pensively through the windows below. Overhead, the sky is relatively quiet, criss-crossed by jet contrails that have warned off the helicopters for the local television stations. And everyone waits. An extraterrestrial craft has landed on Earth. If this were a movie, a diminutive gray being would exit it and say, "Take me to your leader."

But this isn't a movie. It isn't a book. It's real. With all eyes on the craft, a glowing crack appears on its flawless surface, revealing what is clearly a door and a ramp. The crowd holds its collective breath and sees emerging a . . .

. . . a what? That's really the question behind this book. What will we see when we encounter our first example of intelligent extraterrestrial life? Will it be a humanoid Gray? Will it be Adamski's Space Brothers? Will it be Jabba the Hutt, ET, or Spock?

Of course it won't. Nor can I tell you what it will be. It will be Alien, for sure. To remind us of Haldane's famed quote, it will not only be weirder than we imagine, it may well be weirder than we *can* imagine. But we'll try.

The Alien will be intelligent. It will have technology that surpasses our own. It will have limbs to manipulate the world around it and it won't be a water breather. It will almost certainly be able to see light in the spectrum of its parent star. It won't be able to breed with humans, nor will it likely be able to eat and digest Earth food. It will be a curious being and most likely it will be one that consumes resources, perhaps resources found here on Earth.

The Alien is likely to be carbon based and may well use oxygen in its respiration. It likely won't be a traditional plant, although a photosynthetic animal is certainly possible.

But it will be a kindred spirit, thinking as well as a man, but different from a man. It will be a fellow traveler in this universe. It will be an ally and an enemy. It will be an opportunity to learn and to teach.

The distance between stars is large and it may be difficult to travel between them. Perhaps our first encounter with an Alien species will not be a landing on the White House lawn, but hidden in the hiss of a radio transmission. Perhaps the closest we will ever come to Aliens is to view them in their video signals. Somehow I find it hard to believe that if we ever discover that we have Alien neighbors that we won't be drawn to visit them. So that wavering transmission from the depths of interstellar space might one day evolve into a visit to the neighbors.

Or we could be alone in the galaxy, or at least alone enough that encountering an Alien race is improbable in the next thousand years. The Drake equation with modern numbers in it suggests that there may not be many technologically advanced intelligent species out there. Somehow that would be a shame, a terrible waste of space, and yet a wonderful opportunity for humanity. As the planet hunters find Earth-like planets, humans would have places to go, new vistas to explore.

We won't know the answer until we find an Alien. Until that happens, mankind will continue to turn some of our best scientific minds to the question. But in the meantime, we will have to do what we have always done, which is to turn our gaze upward and dream. While we wait, maybe it's best that we should take the advice from the classic movie *Thing from Another World*.

Watch the skies. Everywhere, keep looking. Watch the skies . . .

SUGGESTED READING

General

Steven J. Dick, *The Biological Universe: The Twentieth-Century Extraterrestrial Life Debate and the Limits of Science,* Cambridge University Press, Cambridge, UK, 1996.

Steven J. Dick, *Life on Other Worlds: The Twentieth-Century Extraterrestrial Life Debate,* Cambridge University Press, Cambridge, UK, 1998.

Early Aliens

Robert Crosley, *Imagining Mars: A Literary History,* Wesleyan, New York, 2011.

Michael J. Crowe, *The Extraterrestrial Life Debate, 1750–1900,* Dover, Cambridge, UK, 2011.

Michael J. Crowe, *The Extraterrestrial Life Debate, Antiquity to 1915: A Source Book,* University of Notre Dame Press, Notre Dame, IN, 2008.

UFOs

George Adamski, *Pioneers of Space: A Trip to the Moon, Mars and Venus,* Leonard-Freefield, Los Angeles, 1949.

George Adamski, *Inside the Space Ships,* Abelard-Schuman, New York, 1955.

George Adamski, Leslie Desmond, *The Flying Saucers Have Landed,* Werner-Laurie, Newcastle, DE 1953.

Kenneth Arnold, *The Coming of the Flying Saucers,* privately published, 1952.

Charles Berlitz, William L. Moore, *The Roswell Incident,* Grosset & Dunlap, New York, 1980.

Susan Clancy, *Abducted: How People Came to Believe They Were Abducted by Aliens,* Harvard University Press, Cambridge, MA, 2007.

Jodi Dean, *Aliens in America: Conspiracy Cultures from Outerspace to Cyberspace,* Cornell University Press, Ithaca, NY, 1998.

Stanton T. Friedman, Kathleen Marden, *Captured: The Betty and Barney Hill UFO Experience,* New Page Books, Pompton Plains, NJ, 2007.

John Fuller, *The Interrupted Journey: Two Lost Hours "Aboard a Flying Saucer,"* Dial Press, New York, 1966.

John Moffitt, *Picturing Extraterrestrials: Alien Images in Modern Mass Culture*, Prometheus Press, Amherst, NY, 2003.

Curtis Peebles, *Watch the Skies!*, Berkeley, New York, 1995.

Carl Sagan, *The Demon-Haunted World: Science as a Candle in the Dark*, Ballantine Books, New York, 1997.

Dugald A. Steer, *Alienology*, Candlewick, Somerville, MA, 2010. For children, ages 8–12.

Erich von Däniken, *Chariots of the Gods*, Bantam Books, New York, 1972.

Fiction

Wayne Douglas Barlowe, Ian Summers, Beth Meacham, *Barlowe's Guide to Extraterrestrials*, Workman Publishing, New York, 1987.

Patricia Monk, *Alien Theory: The Alien as Archetype in the Science Fiction Short Story*, Scarecrow Press, New York, 2006.

Life on Earth

Stephen Jay Gould, *Wonderful Life: The Burgess Shale and the Nature of History*, Norton, New York, 1989.

Angeles Gavira Guerrero, Peter Frances, *Prehistoric Life: The Definitive Visual History of Life on Earth*, Dorling Kindersley, New York, 2009.

Tim Haines, Paul Chambers, *The Complete Guide to Prehistoric Life*, Firefly Books, Ontario, Canada, 2006.

Simon Conway Morris, *The Crucible of Creation: The Burgess Shale and the Rise of Animals*, Oxford University Press, Oxford, UK, 1998.

Biochemistry

Jeffrey Bennett, Seth Shostak, *Life in the Universe*, 2nd ed., Addison-Wesley, Boston, 2007.

Iain Gilmour, Mark A. Sephton, *An Introduction to Astrobiology*, Cambridge University Press, Cambridge, UK, 2003.

National Research Council, *The Limits of Organic Life in Planetary Systems*, http://www .nap.edu/catalog/11919.html

Clifford Pickover, *The Science of Aliens*, Basic Books, New York, 1999.

Kevin W. Plaxco, Michael Gross, *Astrobiology: A Brief Introduction*, 2nd ed., Johns Hopkins University Press, Baltimore, 2011.

Erwin Schrodinger, *What Is Life?*, Cambridge University Press, Cambridge, UK, 1992.

SETI

Albert Harrison, *After Contact: The Human Response to Extraterrestrial Life*, Basic Books, New York, 2002.

Marc Kaufman, *First Contact: Scientific Breakthroughs in the Hunt for Life Beyond Earth*, Simon and Schuster, New York, 2011.

Seth Shostak, *Confessions of an Alien Hunter: A Scientist's Search for Extraterrestrial Intelligence*, National Geographic, Washington, DC, 2009.

Seth Shostak, *Sharing the Universe: Perspectives on Extraterrestrial Life*, Berkeley Hills Books, New York, 1998.

H. Paul Shuch, *Searching for Extraterrestrial Intelligence: SETI Past, Present and Future*, Springer, Little Ferry, NJ, 2011.

Peter Ward, Donald Brownlee, *Rare Earth: Why Complex Life Is Uncommon in the Universe*, Springer, New York, 2003.

Steven Webb, *If the Universe Is Teeming with Aliens . . . Where Is Everybody? Fifty Solutions to Fermi's Paradox and the Problem of Extraterrestrial Life*, Springer, New York, 2002.

INDEX

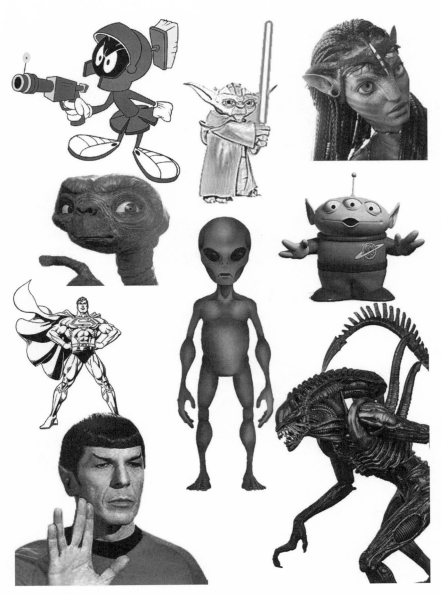

KEY TO ICONIC ALIENS *Clockwise from top left*: Marvin the Martian (from *Haredevil Hare*, Warner Brothers); Yoda (from *Star Wars*, Lucas Films); Neytiri (from *Avatar*, 20th Century Fox); Little Green Men (from *Toy Story*, Pixar/Disney); Alien (from *Alien*, 20th Century Fox); Spock (from *Star Trek*, Desilu Productions); Superman (from *Superman*, National Allied Publications); E.T. (from *E.T. the Extra-Terrestrial*, Universal Pictures); and (*center*) a stereotypical Gray (multiple sources as discussed in the text).